AF806058

Computational Music Science

Series Editors

Guerino Mazzola
Moreno Andreatta

For further volumes:
http://www.springer.com/series/8349

Guerino Mazzola

Musical Performance

A Comprehensive Approach: Theory, Analytical Tools, and Case Studies

Springer

Prof. Dr. Guerino Mazzola
Ferguson Hall 164
School of Music
University of Minnesota
Minneapolis, MN 55455
USA
mazzola@umn.edu

ISSN 1868-0305 e-ISSN 1868-0313
ISBN 978-3-642-11837-1 e-ISBN 978-3-642-11838-8
DOI 10.1007/978-3-642-11838-8
Springer Heidelberg Dordrecht London New York

© Springer-Verlag Berlin Heidelberg 2011
This work is subject to copyright. All rights are reserved, whether the whole or part of the material is concerned, specifically the rights of translation, reprinting, reuse of illustrations, recitation, broadcasting, reproduction on microfilm or in any other way, and storage in data banks. Duplication of this publication or parts thereof is permitted only under the provisions of the German Copyright Law of September 9, 1965, in its current version, and permission for use must always be obtained from Springer. Violations are liable to prosecution under the German Copyright Law.
The use of general descriptive names, registered names, trademarks, etc. in this publication does not imply, even in the absence of a specific statement, that such names are exempt from the relevant protective laws and regulations and therefore free for general use.

Cover design: WMXDesign GmbH

Printed on acid-free paper

Springer is part of Springer Science+Business Media (www.springer.com)

For
Christina
Whose Unique
Performance of Life
Blossomed in My Music

Preface

Musical performance is probably the most complex field of music. It comprises the study of a composition, understanding its expression in terms of rationales stemming from analysis, emotion, and gesture, and then its transformation into physical, i.e. acoustical and embodied reality. Performance communicates its contents and does so in the rhetorical shaping of abstract score data. It comprises a creative interpretation that turns formulaic facts into dramatic movements of human cognition.

Performance is complex, but not necessarily more difficult than its ingredients, such as compositional sophistication or music theory. Its critical quality is the balanced combination of those rationales, shaping strategies, and instrumental virtuosity. Combining these components in a creative way turns out to be a mix of knowledge and mastery, which is everything but straightforward and more resembles the cooking of a delicate recipe than a rational procedure.

Therefore, a comprehensive treatise of musical performance is a difficult business that cannot be achieved as a simple sum of its constituents, but must focus on the interplay of all named aspects of music. Moreover, including analytical tools and case studies turns this project into a demanding enterprise that deals with detailed construction modes and experimental setups of concrete performances—all the more since this book is the first one aiming at such comprehensive coverage of the topic. The extension of the matter reaches from musicological and philosophical aspects studied for example by Daniel Gottlob Türk or Theodor Wiesengrund Adorno [21], to empirical and scientific performance research that germinated with Johann Hohlfeld's *Fantasiermaschine* and was brought to a first florescence with the support of modern computer technology by Johan Sundberg and collaborators at the Kungliga Tekniska Högskolan (KTH) Stockholm [132].

In view of this delicate situation, we are happy that this book could be written with the ideal background and testbed of a course delivered to music performance students, who inevitably want to approach the subject from their concrete situation when performing a musical composition and shaping musical expression under realistic conditions. It is especially this context that makes clear that education of musical performers should not be restricted to the canonical practice that is oriented towards a solid knowledge of the repertory and its technical mastery. A well-educated musician must know for what rationale his/her performance is shaped in one or another way, and which are the parameters that are responsible for the performance's specific qualities.

For this reason it is definitely not sufficient to teach and learn performance according to the old-fashioned model of intuitive imitation of the teacher's antetype. This one is an undeniably precious component, but it cannot play the role of a reliable and exclusive tool in the understanding of what performance is about. It is not reliable since it dramatically lacks the poetical precision asked for by Adorno's and Walter Benjamin's micrologic of performance. It lacks this precision since, although it has a highly developed consciousness of performance as a whole, it does not explicate its constitutive parts and their interaction.

And one also needs to include a greater variety of criteria, more precisely the reference to analytical, gestural or emotional insights, in order to understand and judge performance as a function of semantic layers of the musical text that underlies a given score. Without such alternatives to intuitive imitation, performance risks being degenerated to arbitrariness and disconnected from what has to be communicated to the audience. This is in no way an attempt to construe a unique ideal view of a musical work, that famous inexistent unicorn of performance. On the contrary, semantics includes an infinite variety of perspectives, and the opening of such richness must be enabled to overcome the plain and ultimately sterile individual taste.

This being said, the present book is all but a complete coverage of the broad knowledge and research that deals with performance, its stylistically differentiated practice, pedagogy, and history. Comprehensiveness simply alludes to a conceptual and methodological architecture that surrounds the essential aspects of performance. This is a significant difference to my book *The Topos of Music* [84], which—at the time it was written—was thought as a rather complete reference to mathematical music theory. Therefore the present book is a first sketch of what the overall field of performance could look like as a modern *scientific field*. We do insist on this specification: not the art of performance, but its science. I have often been asked whether this field would ultimately aim at the elimination of human performers in favor of computer programs that would generate musical performance on suitably driven musical instruments from computer-generated analyses of score data in MIDI or similar formats. The answer is no! and it is so much as scientific poetology would never want to replace the creative act of a poet. Performance theory is about analysis, understanding, and experimental simulation of performance as an intellectual endeavor, not as a genuine artistic activity. Which does not mean that performance theory could not inspire artists to explore new creative approaches.

I am grateful to my students at the School of Music of the University of Minnesota for their inspired, inspiring, and collaborative contributions in the

Fall 2009 class. I am particularly grateful for the contributions of Sara Cowan, I-Yi Pan, Cory J. Renbarger, Florian Thalmann, and Nickolai Zielinski for their collaboration in the production of this book. I am grateful to James Holdman's recording and transcription of Egyptian Maqam music. Also, Lisa R. Rhoades' analysis of Miles Davis' improvisation on the standard *I Thought About You* has opened the perspective of an approach to performance that could help move future research in a fascinating new direction. Emily King has been an invaluable help in transforming my text to a valid English prose. To all of them I owe my deepest gratitude. Once more I was given the pleasure to terminate the writing of this book in my brother's beautiful Vulpera domicile—thank you so much, Silvio! Last but not least, I am pleased to acknowledge the strong and singular support in writing such a demanding treatise by Springer's science editor Stefan Göller.

Vulpera, July 2010 Guerino Mazzola

Contents

Part VII References, Music Examples, Index

Part I

Introduction

1

Introduction and Overview

C'est l'exécution du poème qui est le poème.
Paul Valéry

Musical performance deals with the transformation of a (typically classical Western) score into a physical entity composed of acoustical events and the embodiment of the score's symbols in the musicians' bodies. This is already a strong restriction of musical activities, since it excludes musical utterances that are not generated by scores. For example, oral traditions from different ethnicities and the standard practice of jazz are excluded. Jazz, when based on lead sheet notation, uses strongly reduced information requiring a significant interpretational work in order to represent the concretely played sounds. This means that we conceive a score as containing exhaustive data concerning the sounds to be played. A fortiori, free jazz and any kind of free improvisation are excluded from our discourse since they do not rely upon notated symbols, or at least do not presuppose any detailed written information about what is to be played. In order to give an idea of what may happen to the performative activity in improvised jazz music, we will however give a short account on lead sheet driven improvisation by Miles Davis in section 18.3. We shall make the setup of such score-based performance transformations precise in part II: *Structure Theory*.

Part I: *Introduction and Overview*, presents a short history of performance theory and research. This history is far from exhaustive and will only present those key points that relate to the overall concept and theory of performance. A full-fledged history is still outstanding, but this is a logical consequence of the fact that no comprising concept of a performance *theory* has been developed to date.

In order to develop a comprising theory of performance, we first need to unfold the ontological framework wherein performance takes place. This is exposed after the short history of performance in part I. It is essential to

G. Mazzola, *Musical Performance*, Computational Music Science,
DOI 10.1007/978-3-642-11838-8_1, © Springer-Verlag Berlin Heidelberg 2011

describe this context since performance shares a complex ontology, including the manifold of realities, the stream of communication, the semiotic layering, and the levels of embodiment.

Structure Theory, part II, is the first pillar of performance theory: It describes what kind of structure *is* performance. It does not deal with the rationales of performance that relate to *why* a determined performance structure has been chosen. This latter is the second pillar of the theory and will be dealt with in part III, *Expressive Theory*. The typical example used to distinguish these two perspectives is agogics, i.e. microtiming. Structure theory would describe precisely what *is* agogics, while expressive theory would investigate the analytical, gestural, or emotional reasons for shaping a specific agogics for a given time segment of the score in consideration.

Structure Theory gets off ground with the discussion of the score space, where the note symbols are situated as numerically parametrized points, and the space of physical sounds (possibly extended by parameters for musicians' body motion). The performance transformation is then defined as a specific mathematical map, which can be described by means of a vector field on the score space. This is the canonical generalization of well-known classical structures, such as tempo and intonation curves. Performance fields are precisely what Adorno and Benjamin could have envisaged when conceiving what they called micrologic, had they known about differential calculus. Performance fields combine all shaping aspects of classical note parameters, such as onset time, pitch, loudness, duration, glissando, and crescendo, as well as their mutual influence, for example, if intonation is a function of onset time.

As is known for tempo or intonation, performance is only well defined if one knows where to start. For tempo, the starting point is the onset of the initial notes; for intonation it is the concert pitch. The general theory of this initial data, which is needed for all parameters, is dealt with in the chapter on initial events.

The final chapter of this part deals with the global structure of a performance. This one can be described as a hierarchy of so-called performance cells. Such a cell describes exhaustively all data necessary to perform a given set of notes for a fixed collection of musical parameters. For example, the cell could capture all notes with the usual four parameters—onset, pitch, loudness, and duration—call it $\mathcal{C}_{EHLD}$ (this notation will become clear later). However, it could happen that the onsets of these notes can be calculated without any reference to the other three parameters. This means that we also have a cell $\mathcal{C}_E$ of note onsets only, such that physical onsets may be calculated from this onset cell without any reference to the above four-dimensional cell. It will be shown that this pairing defines a projection $p : \mathcal{C}_{EHLD} \to \mathcal{C}_E$. A performance hierarchy is the general diagram structure of the possible projections between different cells of a specific performance.

Once we have completely described performance as a structure in part II we open the discussion in part III, regarding the strategies that shape the components of performance. The history and theory of performance exhibits three basic types of rationales that give rise to shaping strategies: emotions, gestures, and analyses. We discuss each of these rationales and their representatives in performance research. It turns out that performance shaping is by far most evolved starting from analytical rationales. Emotions as well as gestures are known to be of great importance, but the present state of the art is far from what could be accepted as a scientific theory.

We therefore continue our discussion with a crucial method that connects analysis to performance structure: analytical weights. Intuitively speaking, such a weight attributes to each note a numerical value that corresponds to the relevance of that note with respect to a specific analytical point of view. Using such weights as analytical representatives of the given score structure, we then introduce and classify performance operators, i.e. those strategies that use analytical weights to construe performance fields, cells, and hierarchies. This part is concluded by the presentation of two comprehensive models of performance as well as by a short discussion of aspects of improvisation in jazz performance.

In part IV, *Rubato: Model and Software*, we further develop this methodology by adding to the construction of hierarchies the theory of stemmata, i.e. the genealogical procedure that enables successive refinements of performance hierarchies, a procedure well known from performers' rehearsals that gets off the ground with sight-reading and reaches its goal in the sophisticated version of a performance in concert.

This part is completed by the discussion of string quartet performance theory. This theory was initiated with Ludwig Finscher's habilitation [32] and extended to quantitative estimations of degrees of instrumental freedom necessary for the faithful representation of harmonic and contrapuntal score structures in [75]. It sheds new light onto the performance strategies of string quartets as a function of the rich variety of instrumental parameters of the violin family.

Part IV, *Rubato: Model and Software*, is the third pillar of performance theory and deals with the implementation in the software Rubato of the principles and methods developed in parts II and III. It complements the theoretical approaches on a more experimental and algorithmic level. This is necessary for three reasons:

- First, a theory cannot subsist without corresponding analytical tools capable of testing the viability of theoretical models and analyzing empirically given performances. Performance research is also, and strongly, an experimental science, as it became clear from the very first investigations (the *Fantasiermaschine* mentioned above) in the eighteenth century.
- Second, analytical rationales are by far too complex to be handled without modern information technology. Analytical weights for harmonic, metrical/rhythmical or motivic/melodic score structures need specific software components for their calculation. Calculations by hand would never be sufficiently precise and detailed.

- Third, performance theory also needs an intense pedagogical and aesthetic dialogue with performers, be they students or professionals, a dialogue that can only be based upon experimental evidence. It is essential to be able to discuss performance strategies in detail and based upon an environment that allows for concrete illustration of all kinds of sophisticated manipulations.

The first chapter of this part deals with the definition of the genealogical tree of a successively refined performance. Such a structure is called performance score and has been implemented in the software Rubato [80], [94]. It is a tree of hierarchies, whose leaves represent the parts of the final performance, whereas its root is the origin of the rehearsal process, also called primary mother in our theory. The next chapter describes the strategies responsible for the tree's unfolding by use of performance operators and corresponding analytical trees, also implemented in Rubato.

What follows is a collection of Rubato-related case studies, including an exercise by Carl Czerny, some compositions from Robert Schumann's *Kinderszenen* op.15, and a performance of contrapunctus III from Johann Sebastian Bach's *Kunst der Fuge*. (Sound examples are listed in the part VII). The concluding chapter deals with the question of whether performance and analytical weights can be proved to be correlated in existing performances. The discussion focuses on statistical investigations. We show that the agogical dynamics (microtiming) of Schumann's famous *Träumerei*, as they were measured for 28 famous performances by Bruno Repp [111], are significantly correlated to the analytical weights given from Rubato's motivic, metrical, and harmonic analysis.

Part V, *Inverse Performance*, deals with the difficult question about the variety of rationales that may lead to a given performance. We first discuss the technical setup, which circumscribes such a variety, including state-of-the-art tools for inverse performance research. We then apply this approach to a comparative critique of Vladimir Horowitz's (1974) and Marta Argerich's performances of the *Träumerei*. We conclude this topic by revisiting the old problem of music critique in light of our technically specified approach to inverse performance. Music critique is in fact an instance of inverse performance since it seeks to understand and validate a concert or studio performance as an expression of a given arsenal of background settings for the analytical, gestural, or emotional rationales.

In the concluding part VI, *Epilogue*, we summarize the theory and look into possible future developments.

Sound examples are indicated by the sign ♫ and refer to the list *Music Examples* on page 263.

2

List of Symbols

Tout objet abstrait,
obtenu, par exemple, par thématisation,
est un geste sur un geste,...sur un geste
sur le sensible primitif.
Jean Cavaillès [13, p. 178/9]

Symbol	Meaning	page of first occurrence
$\wp$	Transformation from symbolic reality of a score to physical reality of sounds	28
E	Symbolic/mental time	48
e	Physical time	48
S	Spatial position of a moving car	48
$speed(S)$	Inverse derivative of function $S \mapsto e$	48
$speed(e)$	Derivative of function $e \mapsto S$	48
$speed(E)$	Inverse derivative of function $E \mapsto e$	49
$T(E)$	Tempo function at onset E	49
A	Amplitude of air pressure variation	59
Hz	Hertz: unit of frequency, 1 cycle/second	59
d	Physical duration	60
$w(t)$	Periodic pressure wave in time	59
I	Unit interval of real numbers between 0 and 1	60

G. Mazzola, *Musical Performance*, Computational Music Science,
DOI 10.1007/978-3-642-11838-8_2, © Springer-Verlag Berlin Heidelberg 2011

P	Period of a periodic function of time	59
l	Physical loudness	60
h	Physical pitch	62
Ct	Cent: physical pitch unit (1/100 semitone)	60
$p(t)$	Air pressure at time t	60
t	Physical time	61
o	Octave coefficient	61
q	Fifth coefficient	61
t	Third coefficient	61
H	Symbolic/mental pitch	62
$S(H)$	Intonation: same as tempo, but in pitch	63
$speed(H)$	Inverse derivative of function $H \mapsto h$	62
L	Symbolic/mental loudness	67
$I(L)$	Dynamics: same as tempo, but in loudness	67
$\wp_L$	Performance transformation on L	67
dB	deziBel: unit of physical loudness	67
X_P	Real value of parameter of type P	70
$\wp_{EHL}$	Performance transformation on EHL	70
$\mathbb{R}^{eh}$	Real value of parameter of physical onset e and pitch h	70
$\mathbb{R}^{EH}$	Real vector space for onset E and pitch H	70
$J(\wp)(X)^{-1}\Delta$	Performance field for $\wp$ at symbolic point X	73
$\mathbf{Ts}(X)$	Performance field at symbolic point X	73
$\mathbb{R}^{ED}$	Real vector space for onset E and duration D	78
$\partial T(E,D)$	Parallel articulation field	78
$\mathbf{Ts}(E,D)$	Performance field of articulation	78
D	Symbolic/mental duration	78
G	Symbolic/mental glissando	79
g	Physical glissando	79
$\partial S(H,G)$	Parallel tuning field	79
C	Symbolic/mental crescendo	79
c	Physical crescendo	79
$\partial I(L,C)$	Parallel dynamic field	79

Δ	Constant diagonal field	81
$\mathbf{Ts}_{P.}$	Performance field at paramater sequence $P.$	81
$\int_X \mathbf{Ts}$	Integral curve of field $\mathbf{Ts}$ through the point X	82
$\mathbb{R}^{P.}$	Real vector space for music parameters of sequence $P.$	82
$\mathcal{C}$	Performance cell	83
$\mathcal{D}$	Performance herarchy diagram	89
$\mathbb{R}^{EHLD}$	Real vector space for onset E, pitch H, loudness L, and duration D	89
G	Geometric constraints	127
M	Mechanical constraints	127
$\wp_{score}$	Performance transformation of score symbols	127
$X(G, M)$	Manifold of continuous curves of hand movements given by the geometric and mechanical constraints G and M	129
$\gamma_{\mathrm{Physical}}(t)$	Physical hand movement curve	129
$\gamma_{\mathrm{Symbolic}}(t)$	Symbolic hand movement curve	129
$\mathbb{Z}_{12}$	12 chromatic pitch class group	138
$I_{f_\#/g}$	Pitch inversion between $f_\#$ and g	138
PARA	One of the real parameter vector spaces	149
w	Analytical weight	149
$\mathcal{N}(X)$	Nerve of maximal meter covering of composition X	151
$Sp(x)$	Function: simplex of maximal local meters containing a point x	152
ν_i	Weight distribution strength	153
$w_\epsilon(M)$	ϵ-weight of a motif M	155
T	Tonic function	156
D	Dominant function	156
S	Subdominant function	156
$riem$	Set of Riemann function values	156
$val_{ton,riem}(Ch_i)$	Fuzzy Riemann function value	156

$Q_w(E, D)$	Transformation matrix of generalized tempo operator as a function of weight w	161
$\partial T_w(E, D)$	Tempo field with weight w distortion	160
$L_X(f)$	Lie derivative of function f in direction of field X	162
Λ	Weight on a space for Lie operators	162
$\mathbf{Ts}_w(E, D)$	Articulation field at onset E and duration D	161
Dir	Directional endomorphism for Lie operators	162
$\mathbf{Ts}_{\Lambda,Dir}$	Performance field of Lie type	162
δ	Decoding function in Todd's generic approach	164
γ	Encoding function in Todd's generic approach	164
Π	Performance procedure in Todd's generic approach	164
Ψ	Listening procedure in Todd's generic approach	164
$Z(\partial, \mu)$	Articulation field for harmonic analysis μ	197
K_b	Naradaya-Watson kernel function with width b	219
$b \diamond f$	b-smoothed function f	220
$\hat{b}$	Support function with width b	219
Ω_ω^X	Beran operator yielding the logarithm of tempo in statistics	222
$\wp^{-1}(P)$	Fiber of performance map over performance P	227
M	Model of expressive performance for piano hand	230
P_M	Hand performance function for model M	230
P	Output parameters of performance	230

3

Short History of Performance Theory

At present, we have no comprehensive theory of performance.
Hermann Danuser [21, p. 320]

A first look at the history of performance research seems to confirm Reinhard Kopiez's thesis [62] that there are two unrelated threads: on the one hand the empirical research as typically represented by Carl Seashore's measurements of agogics [62], and on the other the philosophical research as typically represented by Theodor Wiesengrund Adorno's writings [2]. It is true that these threads can be recognized, and we shall give their short history below, but it is erroneous to consider them as being in any contradictory position. The difference of these approaches to performance is this: The empirical research deals with quantitative aspects of performance, with the numerical specification of all the parameters that are necessary to describe the structure of a performance, while the philosophical approach is concerned with the qualitative aspects of performance, with the thoughts that might be expressed through performance, i.e. the semantic content of such expressivity.

The apparent contradiction of these positions stems from the difficult and long unsettled problem of how to connect measurable quantities to thoughtful qualities in the context of a valid theory. This means that the positions are not in contradiction, but only disconnected from each other by a historical lack of links, which could generate quantitative correspondences to semantic qualities. It is also true that empirical measurements could not be compared to precise predictions issued from theoretical models, either because the models were strictly qualitative, or because their predictive power was not precise enough. Finally, the early discussions of performance were not driven by a scientific spirit, but by a pronouncedly practical perspective driven by performance pedagogy.

G. Mazzola, *Musical Performance*, Computational Music Science,
DOI 10.1007/978-3-642-11838-8_3, © Springer-Verlag Berlin Heidelberg 2011

It is one of the main goals of this book's presentation of performance theory to offer a solution to the missing links between quantity and quality, between empirical and philosophical aspects of performance.

3.1 The Philosophical Tradition

For a rather thorough account on the philosophical tradition, we refer to Hermann Danuser's excellent report [21], the only major omission being an adequate account on Hugo Riemann's contribtion to performance theory [115].

Danuser exhibits a triple thematization of reflection on performance. Although one can never claim a disjoint splitting of a history of ideas, we can schematically depict this tripartition as follows: The historical development exhibits a first phase, roughly lasting from Renaissance to 1800, which is defined by the study of performance as pertaining to rhetorical aesthetics. The second phase is highlighted by Gustav Schilling's work *Musikalische Dynamik oder die Lehre vom Vortrag in der Musik* [121] in 1843. It is characterized by a switch from rhetorical aesthetics to the aesthetics of a musical work, a change that was initiated by Beethoven's creation of autonomous compositions, whose individual ideation would no longer be a subordinate instance of a general rhetorically expressive setup. The third phase is inaugurated by Hugo Riemann's work *Der Ausdruck in der Musik* in 1884 and can be characterized by the attempt to view performance as being an expression of analytical insights of the work *qua* meaningful text. Danuser, however, only includes analytical insights as rationales that are responsible for the performative shaping. He does not deal with gestures or emotional rationales, for example.

The first phase of rhetorical expressivity can be initiated with Nikolaus Listenius' *Musica...* [71] in 1537, dealing with the *science of right singing*. It appears parallel to Vasari's invention of the central perspective in painting in the last decade of the 15th century, which introduces the individaul human position as opposed to the view of the world *sub specie aeternitatis*, the traditional Christian perspective of divine eternity. This new approach to performance stresses the individual position of the artist, and introduces human existence as a significant and valid point of view on the universe.

The breakthroughs of the human perspective of performance can be situated with Claudio Monteverdi's work in 1600, where performance is recognized as either expressing affects of the artist's musical person or inducing such in the audience. But the work has no essential individuality in this regard; it represents general categories of affect. These are transmitted in the framework of general rhetorics, more specifically according to the five classical rhetorical principles:

1. inventio (the arguments)
2. dispositio (the articulation)
3. elocutio (the communicative wording of thoughts)

4. memoria (memory for performance)
5. pronuntiatio/actio (the actual physical performance)

from which the last was the most important in rhetorically oriented performance.

Among the pedagogically oriented works on performance, the most prominents are: Johann Joachim Quantz: *Versuch einer Anweisung die Flöte traversière zu spielen* in 1752; Carl Philipp Emanuel Bach: *Versuch über die wahre Art das Clavier zu spielen* in 1753; and Leopold Mozart: *Versuch einer gründlichen Violinschule* in 1756, see also figures 3.1 - 3.3.

Fig. 3.1. J. J. Quantz (1697-1773)

Fig. 3.2. C. Ph. E. Bach (1714-1788)

Fig. 3.3. L. Mozart (1719-1787)

These works as systematic, but not theoretically shaped. They apply the linguistic rhetorical tradition to the shaping of musical rendition. The progress is mainly in the written documentation of a culture of performance that used to be oral only, and which is to our days still strongly determined by oral education in the schools of music and conservatories. For example, Leopold Mozart describes and illustrates with images the different ways to hold the violin, either to lean it in a relaxed way on your shoulder, or else, in order to play more precisely, to fix it more tightly with your chin. And this also in order to create a specific posture for the social context in which performance is taking place. This all is also thought in the spirit of switching from the traditional understanding of instrumental performance as being an imitation of singing to an emancipated understanding of musical instruments. The human voice and language are being progessively relativized and complemented by specific and autonomous cultures of instrumental performance.

The explicitly theoretical discussion of performance appears with the works of Daniel Gottlob Türk: *Klavierschule* [138] in 1789; August Leopold Crelle: *Einiges über musicalischen Audruck und Vortrag* [18] in 1823; Georg Wilhelm Friedrich Hegel: *Aesthetik* [50] in 1835 (see figure 2.4 - 3.6); Carl Czerny: *Klavierschule* [17] in 1840; and Gustav Schilling: *Musikalische Dynamik oder die Lehre vom Vortrage in der Musik* [121] in 1843 (see figure 3.8).

Fig. 3.4. D. G. Türk (1750-1813)

Fig. 3.5. A. L. Crelle (1780-1855)

Fig. 3.6. G. W. F. Hegel (1770-1831)

Türk makes scientific remarks and explicitly addresses his book not only to students, but also to researchers. Crelle's book has a first clear semiotic structure. Performance is viewed as an exact expression of a work's content, as a whole and in parts. Here appears for the first time the idea of an individual and "autonomous" work character, which requires a specific treatment in order to achieve adequate expression of contents. This idea is germinating from Beethoven's creativity as being in opposition to general schemes, and as such requiring a thorough understanding and corresponding expression. Hegel goes one step further and distinguishes reproductive versus creative artistic production. He views performance as a recreation of the improvisational origins, and as such realizing the transformation of the composer's inside to the performative outside. Hegel's insights are astonishingly modern and remind us of Adorno's thoughts in [2]. Adorno views the ideal performance as a presentation of composed material as if it were improvised. We shall see in our discussion of Adorno's gestural approach to performance in section 14.3 that he connects this improvisational flavor to the gestural embodiment, which overrides dead symbols of the score.

Fig. 3.7. Carl Czerny (1791-1857)

Fig. 3.8. Gustav Schilling (1805-1880)

To these first theoretical writings one would add Carl Czerny's *Klavierschule* [17] in 1840 (figure 3.7) and above all Gustav Schilling's *Musikalische Dynamik oder die Lehre vom Vortrage in der Musik* [121] in 1843 (figure 3.8), which follows Hegel's systematization principles from general to particular. According to Schilling, performance has to comply with these three principles:

- objectivity: the reference to a completed composition when transforming it into sounding reality;
- ideality: the role of the performer identifying and conveying the composition's ideas and contents;
- totality: capturing the totality of the ideas as expressed in the score's symbols.

Fig. 3.9. Hugo Riemann (1849-1919)

Fig. 3.10. Heinrich Schenker (1867-1935)

Although the theory is now an explicit topic when thinking of performance, it is still not clear which are the precise rationales of such musical shaping activity. The thematization of such rationales, and specificaly rationales from analytical inspection of musical compositions, was initiated by Hugo Riemann: *Der Ausdruck in der Musik* [114] in 1884 (figure 3.11), then Heinrich Schenker: *Die Kunst des Vortrags* (unpublished) (figure 3.12), and finally Theodor Wiesengrund Adorno: *Der getreue Korrepetitor* [1] in 1963, also in collaboration with Walter Benjamin.

Riemann specifies musical expressivity in performance as understood by the shaping of musical thoughts, the plastic relief of motives and themes, and the transparency of the entire construction of the work of art. He determines explicit rules of dynamics to shape analytical components such as melodies and harmonic modulations. Schenker argues for performance as an expression of analytical facts, introducing what Adorno will call analytical performance. And finally Adorno discusses works of the second Viennese school, for example Webern's six bagatelles for string quartet (we shall discuss these analyses in detail in chapter 15.1). His thesis is that the performative shaping that does not descend to the specification of single notes of a composition is invalid as a performative rule. His rationales are derived from a strictly analytical point of view. Together with Benjamin, he coins the concept of a micrological procedure that penetrates the infinitely precise dimension of performative activity. We shall come back to these ideas in chapter 8, page 73.

Fig. 3.11. Theodor W. Adorno (1905-1980)

Fig. 3.12. Walter Benjamin (1892-1940)

Summarizing his historical trajectory of philosophical performance theory, Danuser (loc. cit.) defines performance analysis (German: "Interpretations-analyse") as being the translation of structural and formal insights (on the conceptual level of semiotic units) into performative shaping. Whereas rhetorical theory (German: "Vortragslehre," theory of artistic rendition) just applied general rules of rhetorics, such as *pronuntiatio*, to works of music, performanc analysis fully recognizes the *specificity* of such works and expresses their contents in the shaping of performance, a methodology that became necessary with Beethoven's work.

3.2 The Empirical Tradition

For this thread of the history of performance theory, we refer to Reinhard Kopiez's discussion in [62]. He stresses the increasing role of the performer as an autonomous and core player in the creation and sounding realization of music, citing Franz Liszt's treatise *De la situation des artistes* (*Über die Stellung der Künstler*) [72] in 1835 (published in *Gazette musicale de Paris*), where Liszt equals composer, instrumental teacher, and performer as three variants of artistic personalities.

The historically traced empirical performance research described by Kopiez—and this is identical with the general history of empirical performance research—mainly deals with machines analyzing and/or synthesizing performance of keyboard music. This is due to the fact that keyboard technology is the only one traditionally accessible to precise measurements.

It is remarkable that empirical performance research goes back to 1745, when first construction plans for measuring key movements of the cembalo were made. In 1752, Johann Hohlfeld constructed a precursor of piano rolls, the *Fantasiermaschine* (figure 3.13).

Fig. 3.13. Johann Hohlfeld (right) and the Klavier-Cembalo, a device built following Hohlfeld's *Fantasiermaschine*.

This machine was presented in 1753 to the Berlin Academy of the Sciences. It used pencils on paper, and ran 3.45 meters in 6.5 minutes. Carl Ph. E. Bach tested the device and approved it. The only sample was burnt in a fire at the academy. However we have a sample built by Jean-Joseph Merlin in London 1780, this one called *Klavier-Cembalo*.

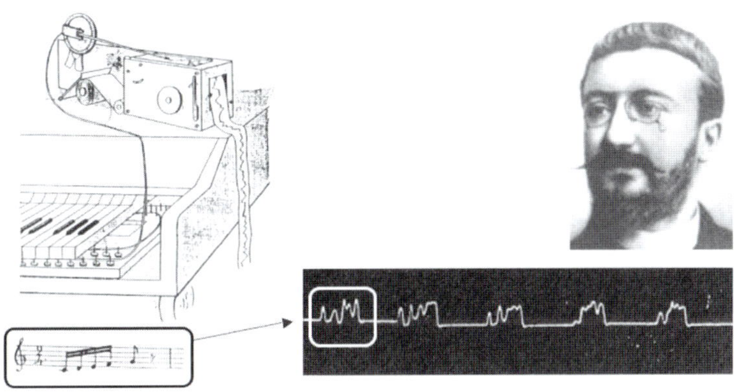

Fig. 3.14. Alfred Binet (right), his machine (left), and the loudness curve measuring several performances of the shown ascending motif.

In 1895, more than a century later, Alfred Binet and Jules Courtier built a machine with rubber air tubes measuring loudness and onset with one millisecond resolution (!), but not measuring pitch (figure 3.14). The results from these measurements show that with increasing tempo the notes are played more legato and less regular in their temporal progression. This is visible from the sequence of five curves shown in figure 3.14.

In 1932, Carl Seashore, the inventor of the Seashore Tests of Musical Ability (his name is translated from the Swedish surname Sjöstrand) invented the *Iowa Piano Camera* at the University of Iowa (figure 3.15).

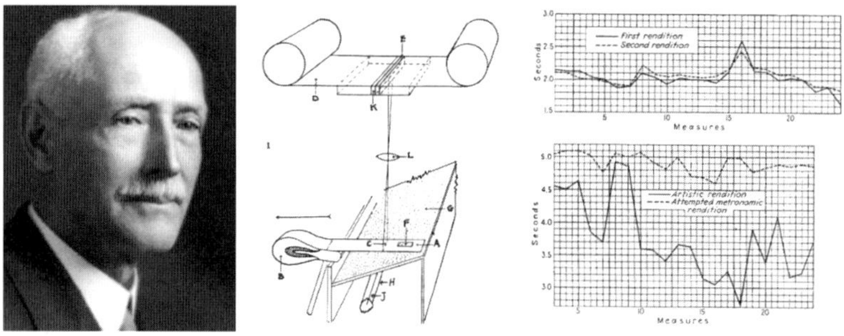

Fig. 3.15. Carl Seashore (left), his Iowa Piano Camera (middle), and the agogics curve (right) showing that agogics is not a random process.

Seashore's results on agogics were groundbreaking. The machine had a resolution of 10 milliseconds and could measure relative note dynamics (in figure 3.15: measured black block lengths from velocity of passing light J through window F on film D) in chords. These measurements proved agogical consistency of different renditions.

After Seashore's death in 1949, empirical performance research paused until the Swedish research team of Erling Bløndal Bengtsson and Alf Gabrielsson started their SYVAR project in 1974 (completed by the studies of L. Henry Shaffer in 1980). Shaffer constructed an optoelectronic machine for a grand piano, feeding the hammer motions to a computer, a precursor of the modern disklavier. Finally, in 1983, the MIDI sound data format [100], [101] was introduced, and the recording of performance became general standard, although Christoph Wagner already had experimented with such devices with accuracy of 1 millisecond in 1974 [145].

Besides this arsenal of devices for the measurement of performances, we build upon the important tradition since 1883 of piano rolls for the analysis of historically important piano performances (see figure 3.16). There are piano rolls recorded by nearly all important pianists of the late nineteenth and early twentieth centuries. Their dynamics and timing is not very precise: They must be considered a simulation rather than an imitation of the performed music (see figure 3.16).

Based upon this data, Rosina Seipp could prove in 1988 [124] that agogics are systematic rather than random (referring to the A-flat major polonaise by Chopin). Around 1992, Hermann Gottschewski [44] had measured a great deal of piano rolls and found rules of agogics. Since 1994, there are MIDI readers for piano rolls by Horst Mohr [98], and also by Zotlan Jánosy and Janos Mácsai,

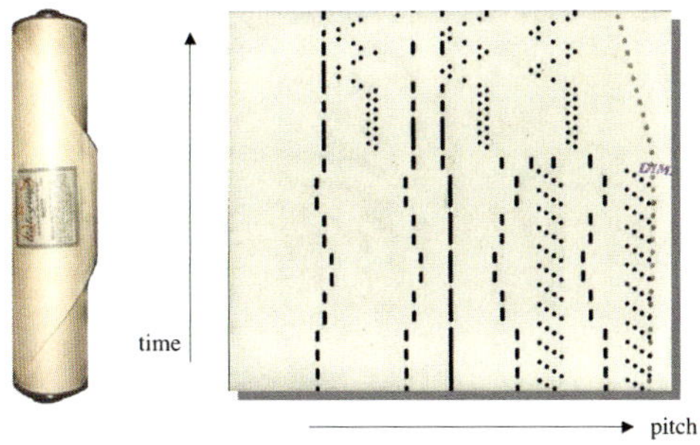

Fig. 3.16. Piano rolls are a historically important source of performance analysis between the last decade of the nineteenth and the first of the twentieth centuries.

but see [62]. Dynamics, however, is still difficult to measure precisely from these piano rolls.

A last method of empirical analysis of performances is the analysis of LPs and CDs. It has a long tradition and is coined "comparative interpretation/performance research". Here are some of the investigations conducted in this field: In 1926, Wilhelm Heinitz [62] compared timings from performances of Wagner's *Meistersinger*. In 1934, Adalbert Kalix [62] investigated ten recordings of Weber's *Freischütz* overture, comparing performance with the score to find out about "musical romanticism." Different compositions, e.g. Beethoven's op.106 *Hammerklavier*, have been analyzed from LPs. Bruno Repp [112] analyzed recordings using FFT for precise onset calculations. In August 2008, Peter Neubäcker published *Melodyne*, a software [105] capable of transforming polyphonic acoustical material into symbolic score format (see figure 3.17). The latter is a real revolution and will change empirical/comparative performance research dramatically. We shall come back to this software in chapter 23.

The two threads of performance research were disjoint from each other in 1992, when my research group started its investigations about a comprehensive performance theory at the Computer Science Department of the University of Zürich. At present, we are able to present a comprehensive theory, uniting the musicological/philosophical and the practical/empirical threads.

Let us conclude this overview with a very short remark about grammatical patterns of performance theory generated by machine-based learning from empirical performance data. We do not, in fact, believe that machine-delegated statistical methods such as neural networks or proper machine learning algorithms for rule learning are of proper scientific value, since when machines learn, we do not. Of course, this is an ideological point of view, but we cannot follow methods that delegate decisions to structured ignorance; understanding cannot be delegated to engineered devices. For example, Gerhard Widmer's elaborate approach [148] starts with a relatively detailed structural analysis of the score, including motives, groupings, etc. It then correlates these structures

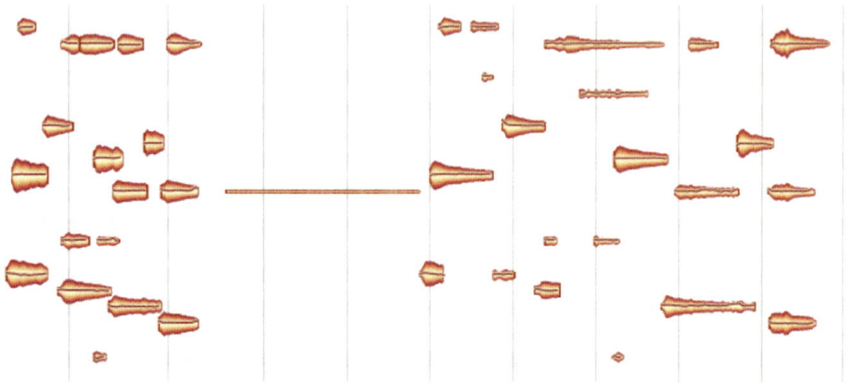

Fig. 3.17. The revolutionary software *Melodyne* is capable of transforming polyphonic acoustical material into symbolic score format.

to empirical performance data, such as dynamics or articulation, in order to apply machine learning algorithms for extrapolation to other scores. It however lacks the conclusions needed for the construction of a systematic performance grammar.

4

Oniontology

Man kann
einen jeden Begriff,
einen jeden Titel,
darunter viele Erkenntnisse gehören,
einen logischen Ort nennen.
Immanuel Kant [57, p. B 324]

The previous, however sketchy, discussion of the history of performance theory has exhibited a split development between philosophical and empirical perspectives. Uniting such categorically different aspects not only is a question of finding a common language, but also requires a unifying epistemological approach. The present state of the art lacks unity on a deeper ontological level. It is, in fact, not even clear which dimensions of musical ontology are involved when dealing with the totality of performance. Evidently, we cannot neglect its deeper semantics, neither can we reduce the topic to purely quantitative, positivistic positions. It turns out that performance is above all complex through its ramified ontological richness.

For this reason, we want to prepend the technically detailed discussion by an ontologically complete picture of music that will enable us to locate performance with respect to all those coordinates known to be determinants of the overall phenomonology of music. This approach will give us the necessary conceptual architecture to unfold a presentation of the subject, which comprises all relevant perspectives and which enables us to interconnect them in the framework of a unified understanding of music.

We shall first describe the general setup of musical ontology and then, in a second movement, investigate the nature of performance within this general context. Although the first topic, the general musical ontology, has been described in a concise way in [91] and in [88], we want to recapitulate it here in

G. Mazzola, *Musical Performance*, Computational Music Science,
DOI 10.1007/978-3-642-11838-8_4, © Springer-Verlag Berlin Heidelberg 2011

order to offer a self-contained text, and also to stress certain aspects in a way as to be more adapted to the context of performance.

The first movement presents what we have coined musical *oniontology.* It is the classical musical ontology that was introduced in [75], comprising the dimensions of realities, communication, and semiotics, but enriched by a fourth dimension, namely embodiment. Since this fourth dimension splits into three layers, the classical ontology is viewed as a triply layered ontology, hence the somewhat fancy name of an "oniontology."

This oniontology presents a topographic landscape of musical ontology; it is a geographic display of localities determined by coordinates as specified from the four dimensions of ontology. In other words, musical oniontology is a conceptual space on which phenomena of musical existence are distributed. This spatial display enables in the second movement to interpret performance as a dynamical process that moves around in the topography, it is an ontological trajectory rather than a constant spot of musical existence. So let us get off the ground with the description of this space of music.

4.1 Realities

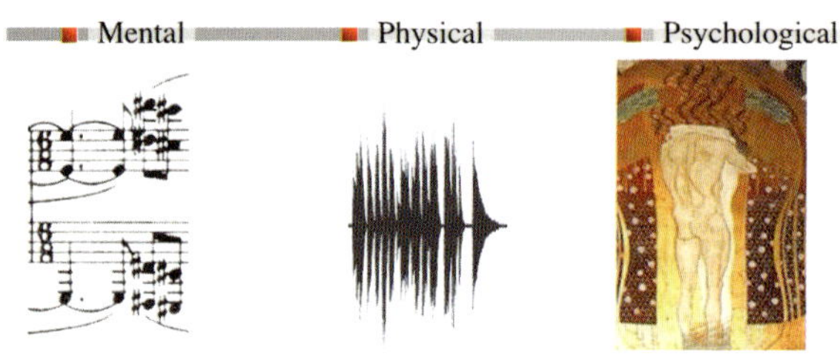

Fig. 4.1. The three fundamental realities of music

This dimension describes the three fundamental values of reality involved in music: physical reality, psychological reality, and mental or symbolic reality. So, acoustical phenomena relate to physics, emotional effects to psychology, and symbolic structures (e.g. mathematical descriptions in music theory) to the mental reality. Observe that the mental reality is not conceived as being a part of the psychological one.

Differentiation of realities is crucial for avoiding widespread misunderstandings about the nature of musical facts.

A representative example of this problem is Fourier's theorem, roughly stating that every periodic function is a unique sum of sinoidal components. Its a priori status is a mental one, a theorem of pure mathematics. In musical acoustics, it is often claimed that—according to Fourier's theorem—a sound "is" composed of "pure" sinoidal partials. However, there is no physical law to support this claim. Without a specific link to physics, Fourier's statement is just one of an infinity of mathematically equivalent orthonormal decompositions based on "pure" functions of completely general character. To give the claim a physical status, it would be necessary to refer to a concrete dynamical system, such as the cochlea of the inner ear, which is physically sensitive to the first seven partials in Fourier's sense.

Methodologically, there is no reason nor is it ontologically possible to reduce one reality to others. For example, it is a logically vicious circle to try to reduce mental reality to physical reality, as it happens in fashionable neuroscience. In fact, explaining mathematical thoughts by neuroscience would mean describing them by chemical and physical processes. But their description would enforce quantum mechanics of chemistry and other basic theories of physical processes. Such descriptions, however, would be based on the complex mathematics of quantum mechanics and therefore generate a vicious circle: explaining maths by maths.

The problem is rather to describe the *transformation rules* from the manifestation of a phenomenon in one reality to its correspondences within the others. To be clear, a neurophysiological transformation ("explanation") of a psychological phenomenon does not, however, conserve the psychological ontology of the phenomenon. The specific phenomenon within the psychological topos *corresponds* to another phenomenon within the physiological topos. But ontologically, the phenomena do not collapse.

4.2 Communication

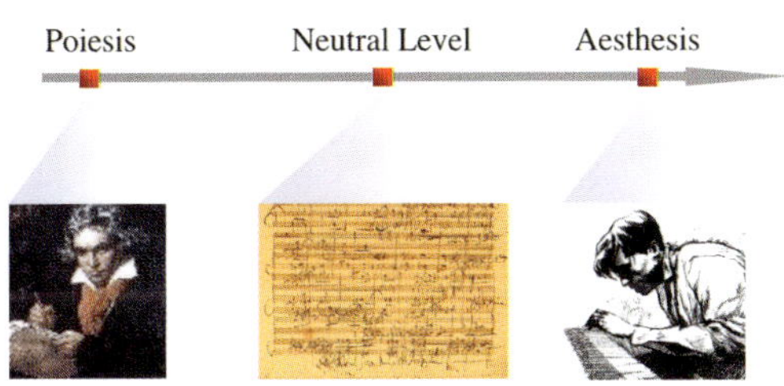

Fig. 4.2. The three stages of communication in music

Following the famous scheme of Jean Molino and Paul Valéry [75], music deals with communication from the first value, the poietical position of the composer or creator, to the creator's work, which is the material essence and output of the second value, called *neutral level* by Molino. The communication, as encoded in the work, targets at the third value: the aesthesic position of the listener, the addressee of the composer's message. Valéry coined the word "aesthesic" to differentiate it from the aesthetical understanding. Aesthesis means perception and can be acoustical, psychological, or analytical, and need not relate to aesthetical evaluation. The aesthesic instance could even be computer software that takes a MIDI file as input and processes an analytical task thereof.

We come back to this dimension in sections 4.11 and 4.12.

4.3 Semiotics

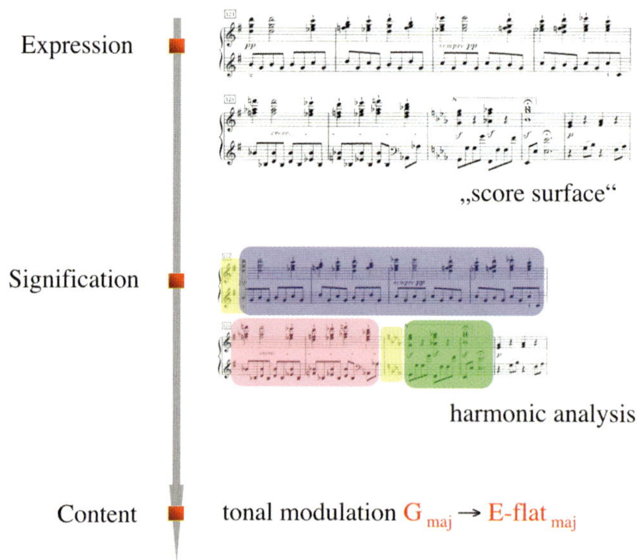

Fig. 4.3. The three positions in musical semiotics

This axis (figure 4.3) comprises all sign-theoretic aspects of music. It is articulated in the three classical constituents of a sign: expression, content, and signification. Expression, the first value on this axis of reality, relates to the surface of a sign, something that stands for the sign's meaning or content. The latter, content, is the second value—the "aliquo" in the classical definition "aliquid stat pro aliquo" ("something stands for something else") of a sign. The third value is the signification part of a sign. It refers to the middle word "stat pro" of the classical definition and explains the way or process engaged for the transfer of the surface value of expression to the "hidden" value of content. For example, when reading the musical expression for a fermata, the reader must invoke a complex machinery to understand the expression, i.e., produce the symbol's content.

The classical three-dimensional cube of musical topography is shown in figure 4.4.

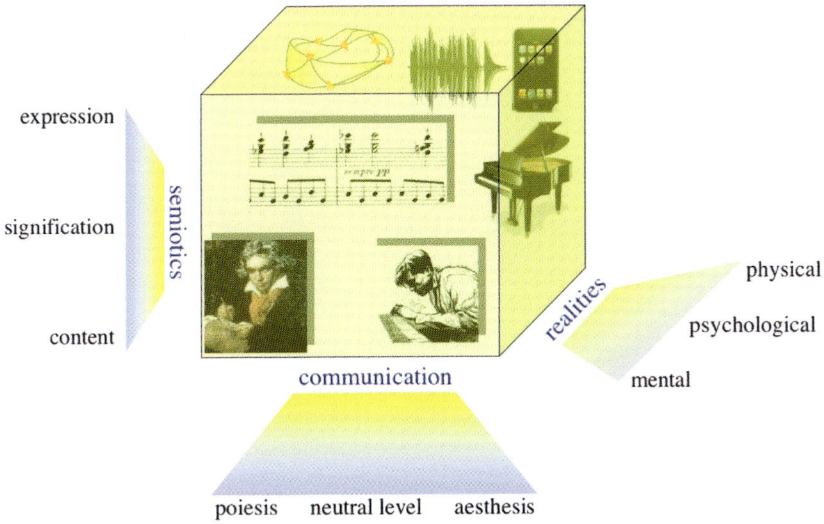

Fig. 4.4. The classical three-dimensional cube of musical ontology.

4.4 Embodiment

The very making of art is a level that is not articulated in that three-dimensional cube of musical ontology. Not one of its 27 ($3 \times 3 \times 3$) positions grasps the gestural aspect of *making* art (and science). The cube, strictly speaking, only deals with the ontology of facts, of "what is the case" in Ludwig Wittgenstein's sense [151]. It does not, however, include the processual level of ontology.

Formally speaking, processes are the diagrams of spaces and transformations that describe the interaction of components of a complex system. We have to differentiate between processes and their products, the output of processual dynamics. Processes are a kind of factory for facts, but not the facts themselves. The processual level is fundamentally different from its output products. Processes and facts are instances of different ontologies.

Going still farther in the initiated direction, processes are also an abstraction from a more basic layer, namely the gestural layer, where all processes and their facts are initiated. Processes are disembodied gestures, reduced to their referential system of transformations.

This entails that a new dimension must be added to the cube of musical ontology. This fourth dimension is coined *dimension of embodiment*. Its three values are: facts, processes, and gestures. They deal with, respectively, these activities: "what is the case," "to refer to," and "to make." In this scheme, the transition from gesture to process is dominated by disembodiment and schematization, whereas the transition from process to facts is dominated by evaluation and dissection (from the relating transformations).

Together with the previous three-dimensional cube of ontology, this fourth dimension creates a four-dimensional cube, which we call the hypercube of musical oniontology. It takes the form of a three-layered onion of gestural, processual, and factual levels of ontology, as shown in figure 4.5.

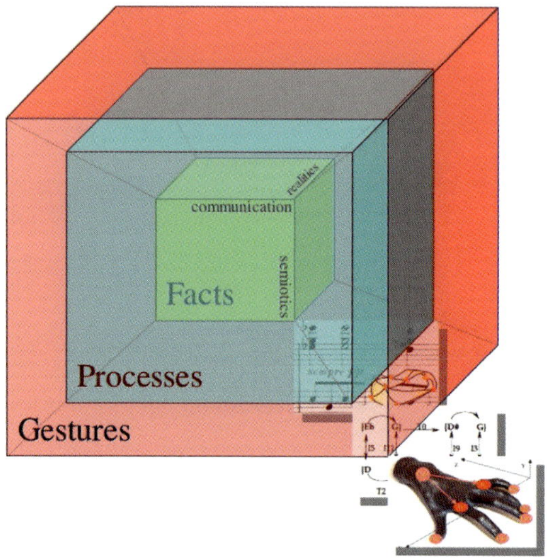

Fig. 4.5. The hypercube of musical oniontology defined by the fourth dimension of embodiment. The graphics illustrate facts (the Möbius strip as a configuration of tonal degrees), processes (a diagram from Lewin's transformational theory), and gestures (a model of the pianist's hand).

4.5 The Baboushka Principle

The above dimensions do not mean that musical ontology is indecomposably inscripted in such coordinates. It mostly happens that the $3 \times 3 \times 3 \times 3$ coordinates are themselves encapsulated subsystems of the same nature. This reiteration of the hypercube's structure is called the Baboushka principle. It does not mean that new dimensions are generated, but that each position in the hypercube can recursively be the compact representation of still a finer hypercube of the same type. Let us make this clear on the two examples of semiotics and communication.

In the semiotic dimension, it is a classical result from Louis Hjelmslev's investigations [52] that the expressive surface of a semiotic system may be a semiotic system in its own. This is the case, for example, in so-called double articulation in language. Here, the words—expressions of the language sign system—are also signs with a graphical expression—the written level of alphabetization—that signifies its acoustical content. This level or semiotic ramification within the expressive level of the top system is called *connotation*. If, on the other end, the content level is itself a semiotic system, the comprising system is called a *metasystem*. And if the middle layer of signification is a semiotic system, the comprising system is called a *motivated semiotics*. It can be shown that music is built from a repeated imbrication of connotative subsystems [83, 84].

In the dimension of musical communication, the overall poietic position may be seen (and this was our example above) as articulated in composer, work (score), and interpreter, whereas this entire communicative unit is the poiesis that generates the neutral level of a performed work (on the acoustical level, say, when the performance is taking place in a concert), which in turn reaches the aesthesic level of the audience.

We leave it to the reader to imagine such Baboushka configurations for the dimensions of realities or embodiment. Hint: Think of the physical reality of a symbol, the symbolic representation of a physical sound, the facticity of a gesture or the processual scheme underlying a gestural utterance, etc.

4.6 Topography of Performance

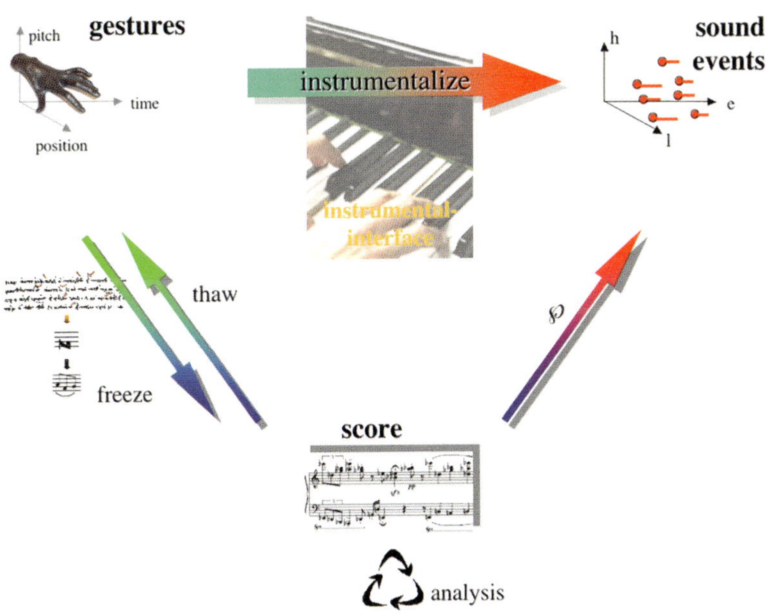

Fig. 4.6. Performance involves the score, possibly its analysis, then the thawing of the score's symbols to gestures that are then transformed into sounds via the instrumental interface.

Performance can now be defined as a transformation of the mental level of the score into a set of sounding/physical events, and this is the type of performance we want to deal with in this book, see figure 4.6. It is crucial to understand this concept as excluding other types of performance not because they are not relevant, but because the chosen type is the perspective that has undergone the most intense and elaborate scientific investigations as revealed in our historical sketch in chapter 3. However, performance involves all oniontological dimensions of music. Above all, the intermediate gestural realization of score symbols, their "thawing to gestures"[1] that act on the instrumental interface and thus generate sounds, plays a major role, but this is—unfortunately—not yet a relevant topic of performance theory. Only the performance transformation $\wp$ from the score to sounds is. Performance theory and practice is, of course, not focused on the mere fact of the transformation $\wp$; the central topic is the investigation and understanding of the transformation's backstage structures. In performance research, these are addressed by the catchword "expressive performance." This somewhat ambiguous concept refers to the communicative process giving rise to $\wp$. As such it starts from the poietic side of the composer and interpreter and is targeted at the aesthesic side of the audience and analyst. This movement is mediated by the performed acoustical and gestural rendition of music.

Such expressivity has two significations:

- It relates to a message that must be transmitted. It expresses a semiotically specified meaning or content. This expressive activity answers to **WHAT** is expressed in performance (see also figures 4.7, 4.8).
- It relates to the means and strategies used to transmit the message to the audience. This is a rhetoric activity and answers the question of **HOW** communication is shaped (see also figure 4.9).

Both semiotic and rhetoric expressivity have to take place and correspond to each other to qualify a performance as being successful: *Good performance communicates contents in an adequate way.*

4.7 Semiotic Expressivity

If a successful expressive shaping of performance is to work, it has to deal first with the semiotic anatomy of the message. Let us first consider the message as it is built from the complex poietic communication unit defined by the trias Composer → Score → Interpreter, which we abbreviate by "CSI." This anatomy is centered around the score's sign expressions and points to a variety of contents distributed among the axes of embodiment and realities (see figure 4.7).

[1] This reverses the creation of a score, which is a kind of "gesture freezing process," as also recognized from its historical development starting with the neumes, those short gestural hints, and ending up with the modern abstract notation.

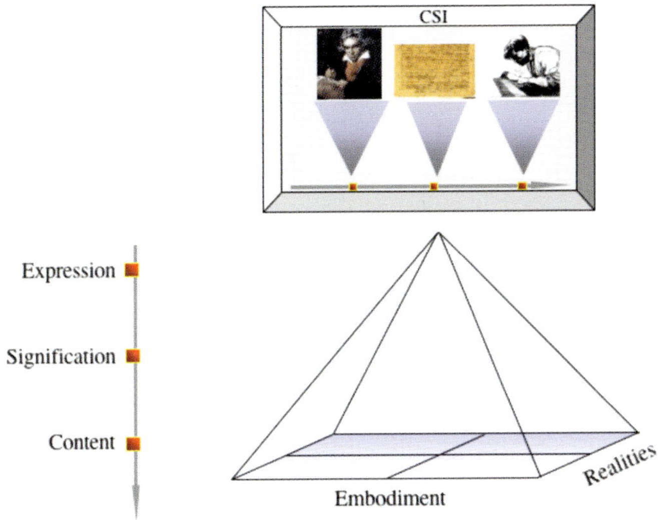

Fig. 4.7. Semiotic expressivity is induced from the communicative unit CSI and targets at the topography of realities and embodiment.

Figure 4.8 shows a selection of possible assignments of score-related contents to the nine positions defined by the two axes of realities and embodiment. This variety makes clear that the semiotic specification of expressivity is a crucial factor that may have strong implicatons on the rhetorical expressivity since its different characters will require different (if adequate) communicative strategies to be successful *qua* communication to the audience.

Let us briefly comment on these examples. In the left column, indexed by the embodiment coordinate "facts," we find the three examples emotions, sound, and harmonic values for psychological, physical, and symbolic/mental facts, respectively. They are just given information, things that are or are not the case. They have no generative infrastructure as such. In the middle column, defined by the embodiment coordinate "processes," we find three examples, one for each coordinate of reality, which carry the character of processes. A psychological drama involves or may involve emotions, but it also has the dynamics of a movement, a diagram of mutual influences and forces acting between different dramatic characters. Similarly, sound generators display a generative process involving different stages and components, such as envelopes, overtones, onset, and duration information (see, for example, the classical description of a sound generator by Max Mathew's Music N processes). On the symbolic level, the well-known motivic work, typically realized by Beethoven, is not a fact, but a complex diagram of motives and their relations of similarity or transformations. On the level of the gestural coordinate, emotions may appear as movements, etymologically relating to e-motion: moving out. Also, Fourier's decomposition of sounds can be expressed as a complex rotational movement of gestures

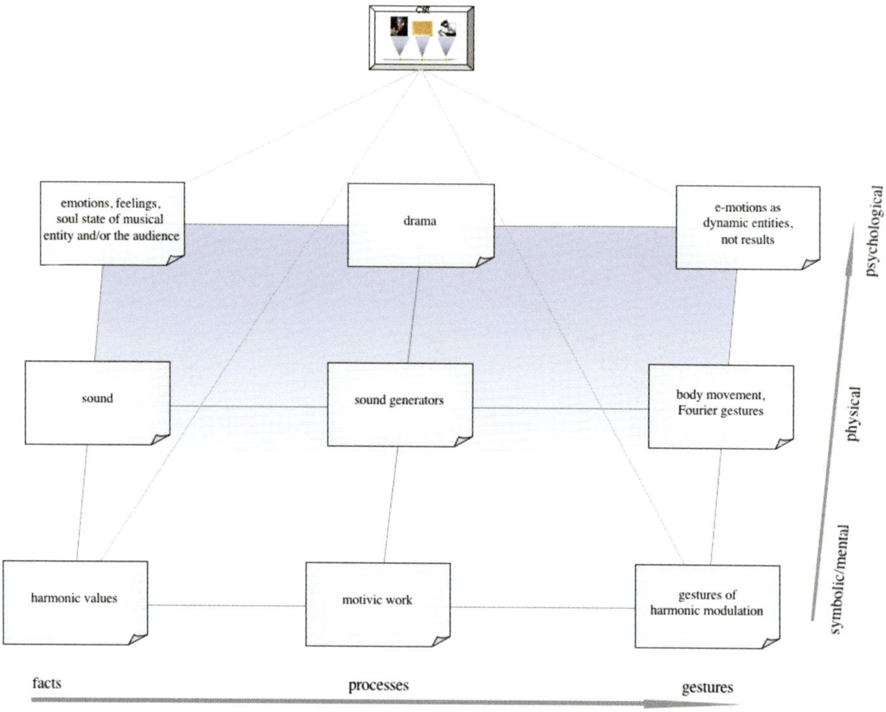

Fig. 4.8. A set of possible message contents relating to the topographic position in he plane of realities and embodiment.

that embody the abstract circle rotation giving rise to sinoidal waves, see [91, chapter 11] for details. Finally, harmonic modulations may be shaped by rich gestural note movements in order to express the dramatic dynamics of tonal transformation, see [93], for example.

4.8 Rhetorical Expressivity

The rhetorical expressivity relates to the quality of the neutral acoustical content generated from the CSI sign by the $\wp$ transformation in order to enable optimal perceptive absorbtion of CSI contents. This means that what is to be communicated arrives in an optimal (though in general not unique) way at the audience level. This quality is a strong function of the part of CSI contents (from the nine above positions) that is being communicated. It is not only a difficult task to realize such a communication for certain parts, given their very nature, but it is also not trivial to give a proof of the communicative success since often there is no additional (meta)communication between audience and artists, either because the composer/interpreter is dead, or because social circumstances prevent such an information exchange.

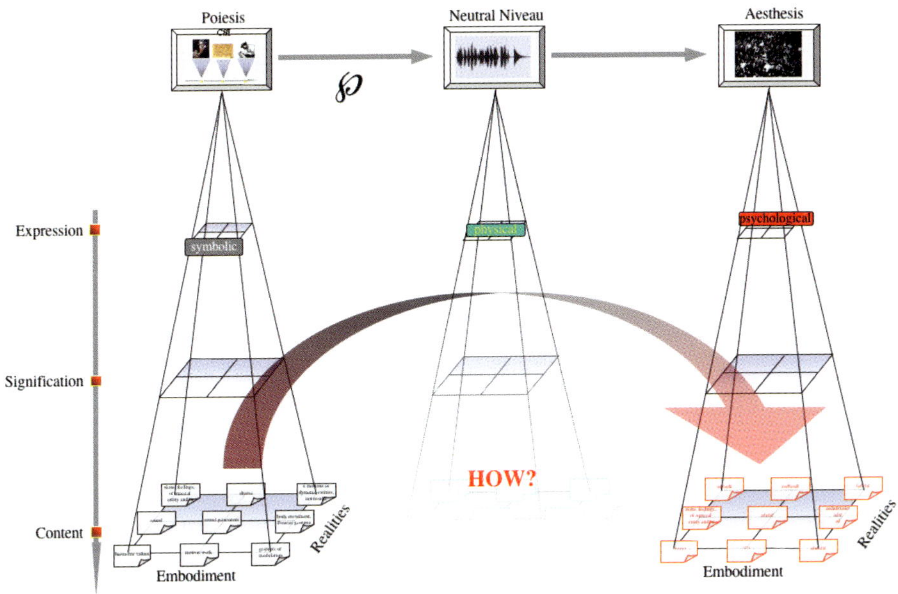

Fig. 4.9. The rhetorical expressivity is a complex communicative process with the bottleneck in the physical (acoustical) level of the performed composition on the neutral niveau.

The composer, the score, and the interpreter (CSI) form a communicative complex in its own, and one whose semiotic dimension is far from homogeneous. This is due to a number of semiotic determinants, in particular those stemming from dia- and synchronic distances, which we discuss in the next section.

4.9 CSI Anatomy

The CSI anatomy is characterized by two separators: Diachronic and synchronic distance between composer and interpreter. These extensions relate to the general fact that a semiotic system expands in space and time; for example, our language has a spatial extension over the territory where it is spoken. Regional differences at the same time refer to this spatialization. Phenomena in a semiotic system that happen simultaneously are called *synchronic*. The second axis of a semiotic system is time: Given a spatial localization, the system's structure may depend upon the moment of time when we observe it. For example, a word has its history, also known as its etymology. The time axis of a semiotic system is called the *diachronic* axis. Figure 4.10 shows the gross space-time distribution of the music system.

Typically, diachronic distance (also called transversal ethnology) is significant if the composer is dead when his/her score is being interpreted, e.g. Beethoven being interpreted by Maurizio Pollini. It is also significant based

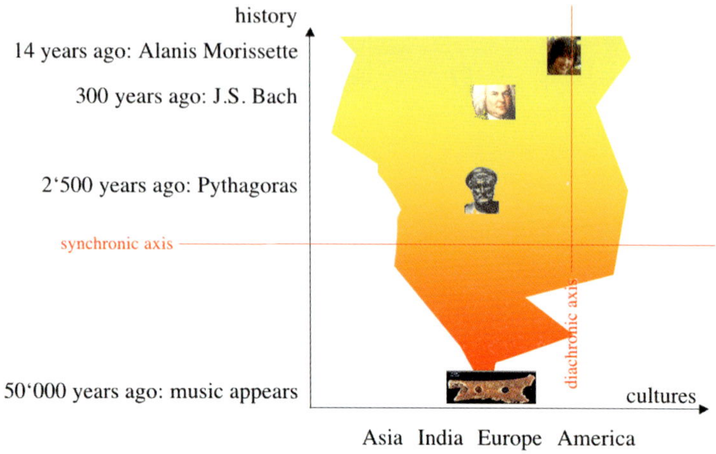

Fig. 4.10. The semiotic system of music evolves in time and space, it is a historical and cultural phenomenon.

on a number of noise factors blurring the transmission through time of the composer's poietic position. Often, such information is missing from the beginning since many composers do not communicate their technical, emotional, philosophical, or religious secrets of composition. It is, for example, common that composers conceal secret messages in their compositions (Alban Berg, Arnold Schönberg, Johann Sebastian Bach, etc.). Such information is usually not transmitted through time, since it is not written down and may only be known by persons that are close to the composer.

Synchronic distance (referring to ethnology) is due to simple socio-cultural separators. The performer is usually not part of the same socio-cultural region as the composer, a fact that might create different understandings of the same composition. But it is also due to the freedom of interpretation. The latter relates to the very nature of communication, which places the aesthesic position symmetrically to the poietic one, i.e., the composer is only understood to be the first interpreter. The interpreter has his/her own rights to understand a composition. This perspective is known in the theory of painting. For example figure 4.11 shows Caspar David Friedrich's *Moonrise over the Sea*, where the observer is integrated in the painting, thematizing the painter's role as being the painting's first observer and interpreter.

It is not mandatory to follow the composer's hints and preferences, in particular if there are fields where such information is simply absent: for example, with Bach's missing dynamic signs, or Beethoven's missing gestural determinants, so dramatically misinterpreted by Glenn Gould in op.57, ♫ 1. If cultural separators include distant ethnics and oral traditions, interpretation may become a dramatic distortion of the composer's intention. For example, if a score is produced that complies with extracultural standards (e.g. fixed pitch as opposed to variable pitch), the interpretation may become unacceptable to the creators.

Fig. 4.11. Caspar David Friedrich, Moonrise over the Sea, 1822.

In the communicative body of CSI, the score is a semiotic bottleneck. It is the neutral reference, but it is a poor information repertory in many regards. The construction of the nine semiotic positions discussed above cannot be completed upon exclusively score-based analysis.

Even with support of the composer (if that is asked for), the composition traced on the score is so rich in details of performative work that the interpreter is forced to recur to knowledge external to the composer's poietics (in particular Kofi Agawu's paratextual attributes: style etc., see [4]) and the score's neutral data. For example, the microtiming (agogics) is virtually never made explicit and has to be shaped by the interpreter alone, using a number of rationales related to the semantic fields under consideration. See also the remarks on tempo in Richard Wagner's book *On Conducting* [144].

4.10 ISC Anatomy

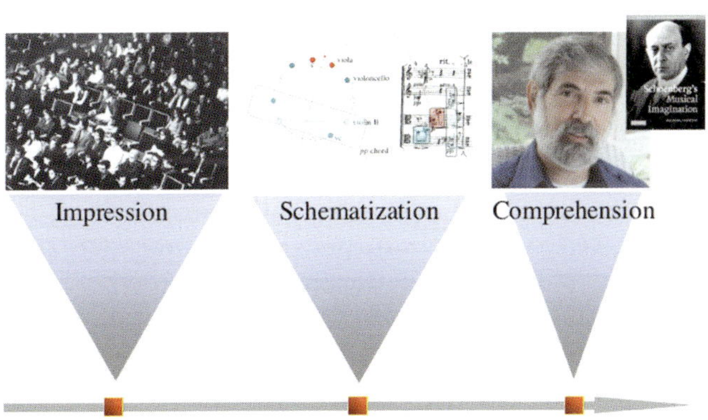

Fig. 4.12. Impression, schematization, and comprehension characterize the communicative unit opposed to the CSI unit on the aesthesic side of performance.

The audience is the aesthesic instance "opposed to" CSI, i.e. lying on the other side of the central position of the work's neutral level. It is, however, somewhat underestimated in its complexity. In fact, aesthesic processing of the neutral level data is everything but elementary and, in particular, involves a symmetric construction with regard to CSI. It is therefore reasonable to call this symmetric communicative configuration ISC = Impression/Schematization/Comprehension. Here is what we intend by these three communicative positions:

- Impression is perceptual as opposed to constructive interpretation,
- the score is substituted by a scheme representing the perceptual body's organization, and
- the poietic creator, the composer, is substituted by the comprehensive force rebuilding the ideas from perception, possibly—but not necessarily—in accordance with the composer's intentions.

This ISC articulation in the audience position is necessary also to explain the huge difference between the understanding of a naive audience as opposed to expert listeners. The idea of the expert listener is typically invoked by authors Ray Jackendoff and Fred Lerdahl of the *Generative Theory of Tonal Music* [53].

This ISC configuration is critial in the judgment of performances. A dramatic situation may, for example, arise from prejudices imported from previous performances to qualify the present one. These prejudices are typically not founded in any logic, but are just there and prevent the audience from dealing with a new type of performance in a fair way.

4.11 Multi-agent Communication

The above architecture of communication in performance is detailed in its poietic and aesthesic dimensions, but it is not detailed with respect to the concrete embodiment of communication by humans and/or machines. It is, in fact, virtually never the case that the communicative agents are in the singular: Several if not many participants are involved in the performative communication of music. The communicative stream is not between two individual instances but is distributed among a number of composers, musicians, audience members, and even—and progressively more so—machines such as computers with interactive programs that may intervene in the shaping of audio output, the algorithmic transformation of compositional input, or the redistribution of musical objects by global social multi-agent music making, as has been proposed by Ali Momeni and coworkers [134]. The musical agents may have multiple communicative interactions, reaching from score-writing composer(s) to performers, improvisers (real-time composers), listeners, and machines that share a variety of roles. It may happen also that such agents are simultaneously involved in different roles, i.e. not only composing, but also listening, reacting, and producing new compositional components.

We therefore propose to enrich Molino's communicative scheme by introducing a multiplicity of poietic, neutral, and aesthesic agents involved in the

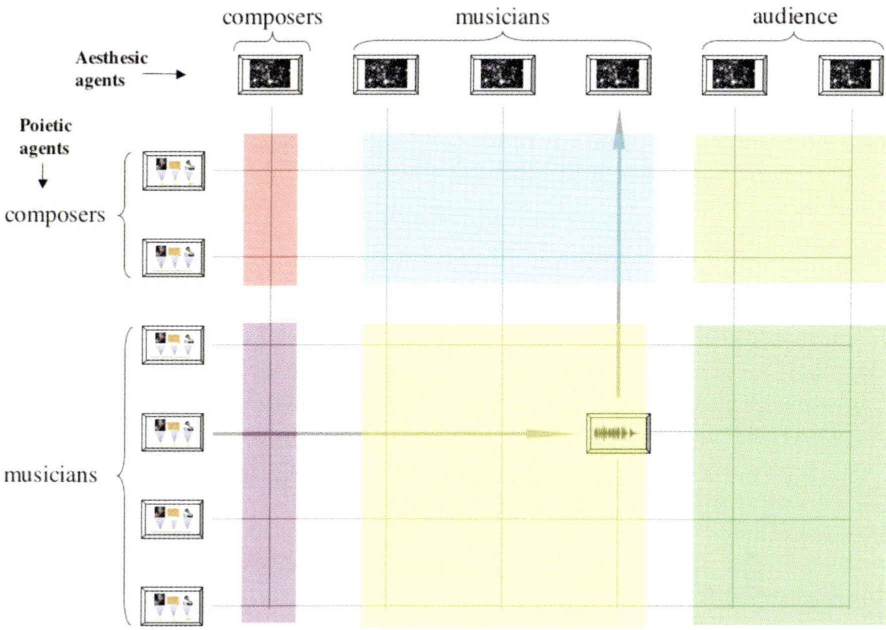

Fig. 4.13. The multi-agent matrix of musical communication. To the left, the rows of poietic agents; on top, the columns of aesthesic agents. For a poietic and an aesthesic agent, we have the corresponding neutral niveau instance.

sonic realization of a musical composition: a *multi-agent communication matrix*. It consists of a series of poietic agents $P(1), P(2), \ldots P(S)$ and a series of aesthesic agents $A(1), A(2), \ldots A(T)$, which are connected to each other by neutral niveaus $N(k,l)$ from $P(k)$ to $A(l)$ for certain pairs. It is not excluded that $P(k) = A(l)$, i.e. the same agent may be poietic and aesthesic! This is the case for improvisers, for example. But we position any such agent in the poietic row position or the aesthesic column position, according to its communicative roles (see figure 4.13).

The figure shows different functions of such agents: poietic composers or musicians and aesthesic composers, musicians, or audience. So, for example, a poietic composer communicates to an aesthesic musican via the neutral niveau of the written score. This is one of the classical relations. But a musician may also act poietically upon a composer, such as when an improvised musical structure is inserted into the composition that a composer is writing. And here, the composer might be identical to the musician in the sense that the composer acts as a musician and then processes the played music in his/her compositional creation. This is a frequent relation in jazz, but also in classical composition, where the composer switches roles during the creative process. In improvised contexts, the communicative relation from musician to musican is standard, and in computer-aided composition, the computer might act upon the human composer in a poietic role.

The overall image of this network is a global field of agents interacting in possibly changing roles and thereby blurring the rigid classical image of a unidirectional communication from creative composers to passive performers and listeners. The musical creation in performance is a network with loops and distributed roles.

4.12 The Performer's Balanced Dancing Presence

Although we shall develop the technical details of structure and rationales of performance later in this book, it seems important to draw a compact image of the existential focus, which is embraced by an engaged performer. We do so for two reasons: first, we would like to sketch an aspect of performance that is vital to the understanding of a performer as a living artist, and second, we want to give a first hint to those performers who rightly cannot be happy with a merely technical understanding of their artistic life. It is only a germinal hint, since the existential shape of a performing artist is far from being understood, and it is not the target of this book to unfold this deep topic. Let us briefly digress on the very difficulty of such an enterprise and have a look at the philosophy of dance as it has been addressed by Paul Valéry in [143].

A dancer is a performer *par excellence*, since the dancer's score, written in Laban notation or any other dance score language, is known to be a poor reference to what dance is when it happens. This has been analyzed in detail in [49]. Valéry, in his treatise, recognizes that dance is more action than every other art, that it is an art of time, and that it is an art in which the artist's life is taken in its full extension as the dancer's body is fully engaged in the unfolding of dance. But then, instead of proceeding to a valid definition of dance (what he explicitly wants to achieve), he looks into the philosopher's mirror and recalls that Aristotle, Nietzsche, and in fact all philosophers are dancing with their words and thoughts. What a pirouette of thought: Valéry, instead of writing a philosophy of dance, makes philosophy dance. This dance of philosophy seems to miss the point: You usually do not represent knowledge about an object by making it. Valéry's approach does not give the expected definition of the concept of dance. However, he says that dance, as an art of time, pertains to a fundamental quality of human existence, namely time. And he recalls that Saint Augustin admitted that he knows what time is, but when asked about time cannot define it. Dance for Valéry is a similar phenomenon: Impossible to define, you have to live it, and he lets us know that thinking is a way of dancing.

Although this insight is precious, we cannot accept the answer "Let's dance, and you will know!" to the question "What is dance?" Of course, the answer confirms that famous saying that dancing is a way of thinking. And for our concerns it is also true that performing music is a way of thinking. But here lies the problem: If it is a way of thinking, in what sense is it a special way, and different from other ways of thinking? The performer thinks music in very specific ways, radically different from the way a music theorist thinks music—see Martin Puttke's paper "Learning to dance means learning to think!" in [110]. So let us try to describe those characteristic coordinates of this way of thinking music.

It is a logical necessity to locate the performer's (oni)ontology in the framework of multi-agent communication since a performer should focus on communication among *all* agents. A performing musician cannot limit his/her interaction to a unidirectional messaging from composer to audience. This is an outdated casting of performative creativity as slavish service for the ingenious composer, an all-too-narrow perspective propagated by Arnold Schönberg and resonating in a casted performance education that produces only robots, not musicians. It is a sad fact that even in jazz school education, the slavish messaging of jazz on the basis of lead sheet changes replaces creative interaction of improvisers and disseminates that "whitened jazz" catechism, against which—apart from more general social and political motives—free jazz in the 1960s of the 20th century was also rebelling.

Within the three dimensions of musical oniontology that complement the communicative dimension, namely realities, embodiment, and semiotics, the performer realizes a crossing of a singular type. In what follows, we stress the characteristic features (which does not mean that other aspects are absent).

The most characteristic feature on the axis of realities is the interaction of two bodies: the musician's body and the body of physical sounds. Their interaction is generated on the interface of the musical instrument, whose bodily manipulation produces the music's sounds. For an acting performer, this coupling of bodies is the core neutral niveau. All other levels of neutrality might be implied or subsumed, but this one is the manifest neutral building block.

On the axis of embodiment, corresponding to the reality of instrumentally interacting bodies, the performer's focus is on gestures. It is these gestures that are communicating musical formulas or processes. It is the highest quality of musical expressivity to deploy compact musical formulas into gestures. Gestural embodiment does not populate given spaces but creates them, defines their extension and thereby enables fellow musicians' gestures to resonate with one's gestures in these shared spaces. "Understanding is catching the gesture and being able to continue." This deep insight by the French philosopher and mathematician Jean Cavaillès [13] is what happens in the gestural interaction among performers: Their gesturally spaced vibration is what drives their bodies to move and to shape the "body of time."

On the axis of semiotics, the (successful) performer organizes the future of the music being performed with reference to the music's past. The meaning of the music played to this present moment is connected to the shaping of the meaning of the next musical signification in a flow of thoughts. This creative transfer is performed by the body's gestural utterance. In order to achieve this in a coherent and persuasive way, we have to identify an environment where such a strong shaping activty is executed. On the level of physical events, we cannot realize such a program, since physical time presence is a real number t_0, past is the interval $t < t_0$, and future is $t > t_0$. Therefore classically conceived physical presence reduces to a single time point, and nothing can really happen in such a vanishing point.

This perspective is not satisfactory from the performer's point of view since the concept of presence in the time-sensitive arts cannot be reduced to zero. We do not embark in a neurophysiological model of performer's presence, because to our knowledge there is no such model. We rather want to postulate such a reality of artistic presence independently of a neurophysiological modeling. The fact is that the time-space of presence in artistic creation and shaping of structures is a huge environment where all the logical decisions upon performative actions are made: gestural strategies, receiving and processing the structures of past musical events, the contributions from other agents in the multi-agent network, the knowledge from the symbolic score and its prefabricated analyses, etc. We need such a time-space that conceptually is independent of the physical time point of presence. Let us call this time-space "imaginary" for two reasons. On the one hand, it is an environment that pertains to the psychological reality and as such is imagined. On the other, and this is a speculative thought, it is known that modern physics—in the research of Stephen Hawking [48] in Big Bang cosmology, and also in contributions of Itzhak Bars [5] to the unification of gravitation and quantum mechanics—has introduced a second time dimension.

In Hawking's model, time is a complex number $t + i.s$, where t is the traditional time value and $i.s$ is the imaginary component. It is not clear at all whether and how this imaginary component could be a part of human consciousness. But we conjecture that our presence of consciousness, where thoughts are built and processed, could happen in that imaginary direction, so that the classical real space-time is complemented by an imaginary time-space defined by imaginary time. If physicists are entitled to introduce new time dimensions, there is no reason to prevent artists from doing so and claiming that creative human consciousness is hosted in such a time-space that is "orthogonal" to the physical one. This means that at any classical physical time t, we would have an entire time-space defined by an imaginary time $i.s$ plus some space coordinates attached to that time.

Based upon such arguments for an extra time dimension, we argue that the concept of presence in time-critical arts requires such an imaginary time and space. It is in this realm where the transitional processing of past music to future music, the planning of gestural strategies, the body-instrument-sound interface are all displayed and organized. It is a quite dramatic change of understanding of what happens in the artistic presence, since it eliminates the mystification of spontaneity and imprevisibility in creative performance. These attributes have had a great influence on the non-understanding of creative performance, from classical music to free jazz. Improvisation, creative performance—all of that has been boiled down to these negative concepts:

Spontaneity and imprevisibility are just negations of any sort of positively defined artistic shaping; they are the "emergent properties" of creativity mysticism, creativity by negation. Telling a musician to be spontaneous is of no help whatsoever: It is just a recommendation to rely on nothing that could be conceived of in the artistic-shaping activity.

Offering to the artist the concept of imaginary time-space is a completely different affair: It opens a huge environment where the artist's consciousness can evolve and construe complex architectures and highways of shaping musical structures. The gestural complex is driven in this time-space; the flow from past to future music that is being played is driven and lived in this imaginary realm. The overall image is drawn in figure 4.14

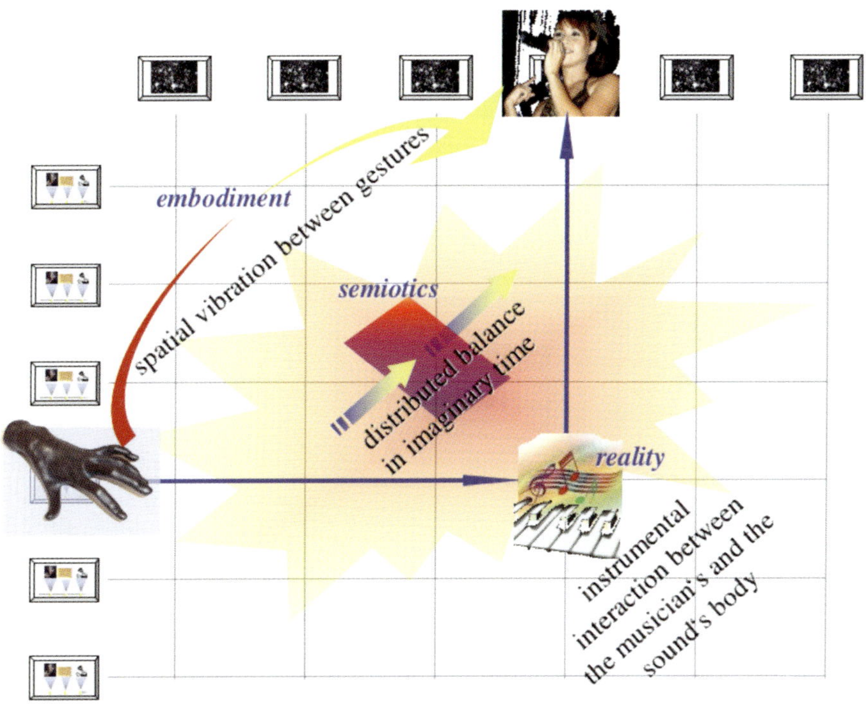

Fig. 4.14. One could conceive musical performance as being the balanced dancing of presence, an existential intensity in three directions, embodiment, reality, and semiotics, of communicative activity.

Let us focus on the semiotic flow process connecting past music instances to the shaping of future ones. This aspect can take place in very different ways. To begin with, the performers may rely on different extensions of the past music's structure. An improviser might only listen to one other fellow musician or he/she might listen to many of them when shaping the next sounds. But these future sounds might also be shaped by a more or less strong reference to pre-conceived structures, independently of the actual performance, such as the reference to a given score that tells me which notes to play next *independently* of how and what has been played to the moment. These factors in the shaping of future sounds are more or less distributed processes: They define an identity of the performed piece of music that is distributed among the multiplicity of agents. This is why we call this process one of a distributed identity. And why we would define the quality of the present performance as being defined by the coherence and strength of this flow as a distributed identity.

Putting the three components of performance together, we see that the pairing of body-instrument-body and of gestural space vibration could be conceived as the aspect of dance. In short, dance would be viewed as a synthesis of body and gesture. A second pairing, body-instrument-body and the flow of structural unfolding, would then be seen as the balance in a bodily realm, a concept that is akin to what classical Greek aesthetics called "kairos"—the perfect balance in the body's dynamics of presence. And finally, the pairing of gesture and structural flow would be understood as a shaping of the body of time, the dynamics within the imaginary time-space that defines our imaginary body of time; we might call this the presence in performance. Putting these three pairings together, one could then conceive *musical performance as being the balanced dancing of presence*, an existential intensity in three directions of communicative activity.

Part II

Structure Theory

5

What Is Structure Theory?

A known mistake is better than an unknown truth.

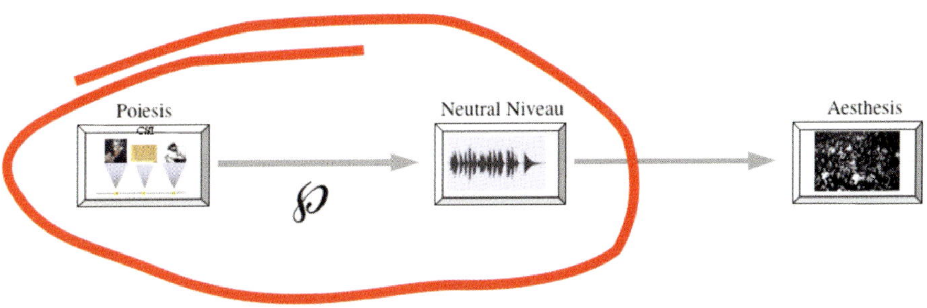

Fig. 5.1. The structure of performance describes the transformation $\wp$.

The oniontological topography developed in the last chapter specifies a communication from the poietic to the neutral position, yielding the performed music as a sounding expression. The delicate point is the technical way of connecting the score information and the surrounding CSI to the sounding output. The solution we have chosen is historically justified in that the score has evolved from the neumatic notation, meaning that the score is a "dance floor of gestures" and therefore relates to real physical events in an abstracted way.

This implies that the score coordinates of onset and pitch are an abstraction from real physical coordinates. The nature of this abstraction is a standardization of time and pitch, subdividing these parameters into quanta: Time is a multiple of standard durations, usually 1/2n (1/2, 1/4, 1/8, etc.) of a measure's duration, and pitch is a multiple of semitones.

G. Mazzola, *Musical Performance*, Computational Music Science,
DOI 10.1007/978-3-642-11838-8_5, © Springer-Verlag Berlin Heidelberg 2011

These standard units are not the physically meant quantities, but can determine such quantities once we have fixed the gauging of time and pitch. So the score space is a symbolic space bearing the potential, when additional information is provided, to generate physical space coordinates.

Such information suggests that the transformation from score to sound is a map on the score space, mapping symbolic events (notes, pauses, bar lines) to physical ones. We call it the *performance transformation* and denote it by

$$\wp : \text{Symbolic Space} \to \text{Physical Space}$$

The precise definition of these spaces is required, of course, and it is by no means simple! This approach is the most precise and efficient one to date. But it has a number of consequences that are as remarkable as difficult for mathematicians, music theorists, and performers. The most difficult one is the fact that such a map is mathematically complex, even in the most common situations. A number of delicate questions arise:

How can we classify such maps? How can we then justify such a mathematical choice of spaces and map properties in terms of common musical situations? Which are the most elementary such maps in music? Is it really necessary to embark in such general math? Isn't music dealing with just the most elementary mathematical situations? Which parts of these maps are connected to reasonable performance parameters, if there are such parameters at all? Etc.

Fig. 5.2. Performance must be described separately for each instrument.

But pure mathematics is not the only problem. The performance transformation takes place in a number of local and global contexts, namely:

- It is a performance with *different instruments*,
- it is a performance of a *patchwork of parts* of a larger composition,
- it is a performance in *different dimensions* (onset, pitch, duration, etc.) which play very different roles—some are dominating, others are more subsidiary.
- And performance is also an *evolutionary process* that evolves from sight reading to a refined performance.

This means that the mathematical description of performance by the mapping $\wp$ must also include these local/global contexts. It will be quite challenging to realize this complex structural description language, but any simplification is impossible, as will be shown later. This is in fact not an artificial blow-up of musical situations, *it is the mucial complexity that enforces the setup.*

Let us briefly discuss these local/global contexts. To begin with, the score is, in general, a union of different instrumental subscores (figure 5.2). Performance structure must start with its specification for each instrument or instrumental group since the parameters that are available depend on the instruments. For example, a violin has glissando and crescendo parameters, which a piano cannot have. Hence performance of these parameters is only reasonable for a violinist, and not for a pianist.

The second localization relates to parts within a specified instrumental score. It is virtually never the case that one performs the score as a whole, but one looks at parts that are chosen for different roles they play in the performance (figure 5.3). For instance, a pianist may choose the left hand versus the right hand part. Also, a phrase or period or even a single chord may be a structural focus of the performative shaping.

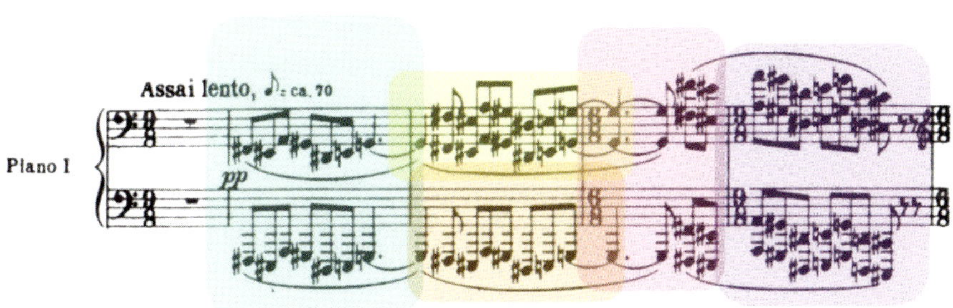

Fig. 5.3. Performance must be shaped for more or less small parts of the instrumental score.

The third localization is concerned with different roles that sound parameters may play in the shaping of performance. For example, duration will depend on the shaping of onsets, but not vice versa. Concretely, this means that when we have a piece played with a given tempo, then we know both about the onsets of notes and their offsets, i.e. the onset of their ending. This means that their duration is in fact determined by the performance of onsets. We shall see that there is an entire *hierarchy of sound parameters in performance.*

The fourth localization relates to the fact that one never performs the valid version of a composition at once, from scratch. Rehearsal is the rule, but not just as a technical necessity. Rehearsal of a performance goes much deeper into understanding the work of art. It is the process of an unfolding logic of performance, a genealogical inheritance process that successively builds more refined structures upon previous ones as in figure 5.4. We shall discuss this *in extenso* in part IV.

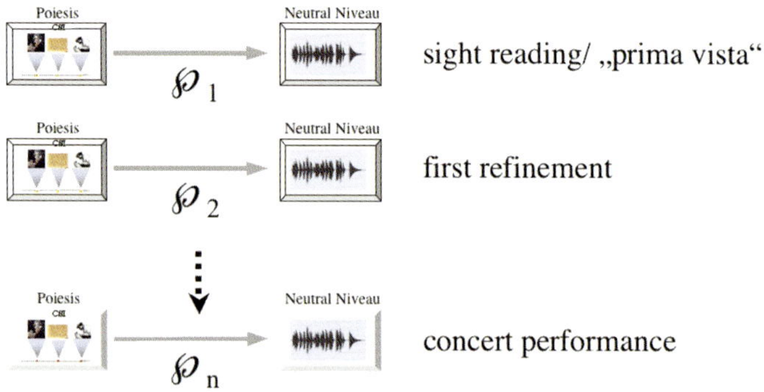

Fig. 5.4. Performance unfolds in a successively refined process of intermediate performances, ending with, say, a valid concert performance.

All these localization aspects add up to a core theory of the structure of performance, namely one that focuses on a usually small part of the score of one instrument, and on a fixed selection of parameters (not necessarily all of those that describe the instrumental sound, but a selection that is defined by an autonomous set of parameters, i.e. independent of other instrumental parameters), and finally on a fixed stage of the genealogical tree.

In what follows, we shall discuss in detail those parameters which are classical and describe the very basis of any structural theory.

6

Tempo Curves

Time stays long enough for anyone who will use it.
Leonardo da Vinci

Tempo deals with the performance of time. There is a symbolic time, as used in the score notation, and the physical time of acoustical sounds, into which the symbolic time is transformed; for example, quarter note time units are transformed into seconds. We denote this transformation by $\wp_E$, E standing for symbolic onset time, whereas e stands for physical onset time (figure 6.1).

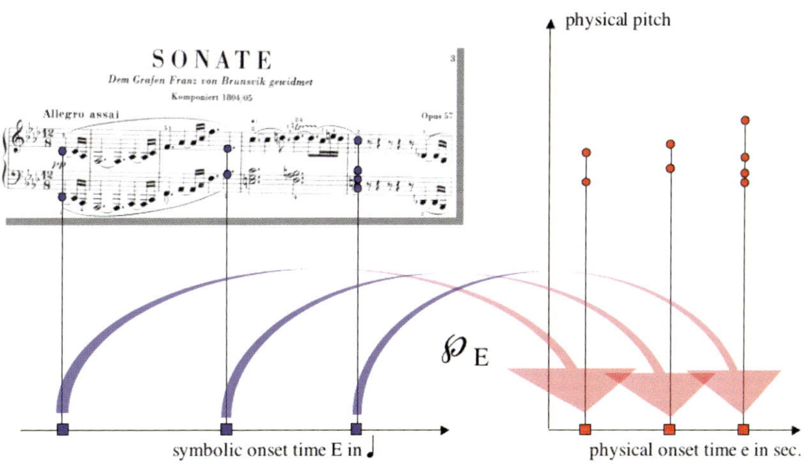

Fig. 6.1. Performance of time transforms symblic time units (typically quarters) into physical time units (typically seconds).

G. Mazzola, *Musical Performance*, Computational Music Science,
DOI 10.1007/978-3-642-11838-8_6, © Springer-Verlag Berlin Heidelberg 2011

What are these spaces of time? Carl Dahlhaus [21, p.53] distinguishes a "potential time" of the score versus the "real time" of performance. He states that "performances are contained in real time, while the score contains potential time." We do not follow this distinction since the symbolic time of the score is as real as the performance time. The difference is not performative reality versus score's non-reality—both are real, but pertain to different realities: the score to the mental reality and the performance to the physical reality in the ontological topography.

Peter Desain and Henkjan Honing [24] argue that symbolic time is discrete, whereas physical time, tempo and expressive timing are continuous. This is erroneous for several reasons:

- Metrical time is infinitely divisible in itself: No positive lower limit for mental durations has ever been envisaged. Mathematically speaking: Metrical time is a topologically dense, not discrete set in the field of real numbers. Hence, any reasonable (more precisely: uniformly continuous) time function from mental/symbolic time E to physical time e can uniquely be extended to a time function on the reals (see [59]). There is no conceptual reason to restrict metrical time to a discrete subdomain of the reals.
- Tempo does not deal with something more continuous than metrical time. It is another concept (see later in this chapter): the inverse differential quotient of a function $E \mapsto e(E)$ between two copies of the real number axis with irreducibly different ontological specifications, namely the musical mental status of the score and the physical status of performed music. Tempo is also constantly present, even within rests, fermatas, or glissandi.

So, summarizing, the space of symbolic and the space of physical time is the real line $\mathbb{R}$ of all real (= decimal) numbers, not just a discrete subset; it is the line of all real values, without holes, like in geometry. So performance of onset time is a map $\wp_E$ from the real time line of symbolic time to that of physical time.

6.1 What is Tempo?

To understand the concept of musical tempo, let us first look at the concept of speed in a physical environment, of a moving car, say (figure 6.2). As the car moves along the road, the driver may ask at any position on the road: What is the time? This is typically the case when the driver has an appointment, where he/she has to be at a certain place at a scheduled time.

This setup implies an interpretation of the car's momentous speed at the spatial position S as follows (figure 6.3). The speed at time e is defined as derivative $speed(e) = dS/de(e)$ of the inverse function $e \mapsto S(e)$. Therefore, using the original function, speed at the position S is the inverse derivative

$$speed(S) = \frac{1}{de/dS(S)}$$

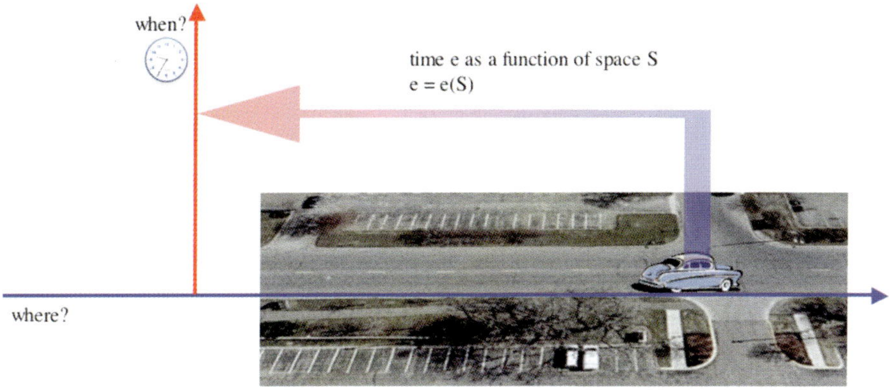

Fig. 6.2. When a car moves through a road, the time can be seen as a function of the position of the car.

as a function of S (!). This is a somehow different point of view since speed is viewed as a function of the position $S(e)$ at time e, and not of time e itself.

Now, the musical analog of this situation views the spatial dimension S being represented by the position of the performer when running though the score. The score takes over the role of the road, the car being replaced by the performer. This is shown in figure 6.4.

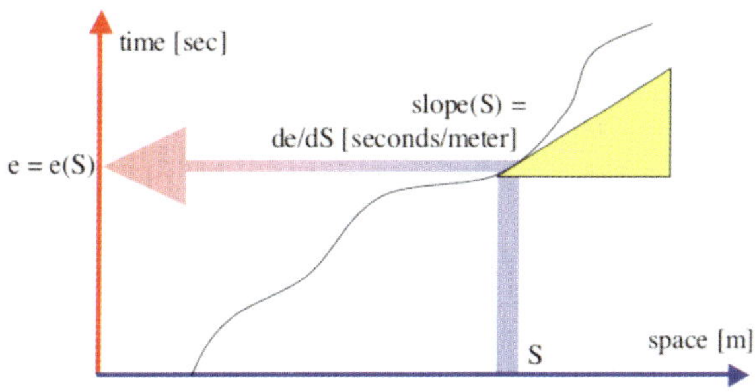

Fig. 6.3. Speed is viewed as the inverse slope of the curve defined by time being a function of space.

In this setup, S being replaced by E, speed takes the form

$$speed(E) = \frac{1}{de/dE(E)}$$

This is precisely what musicians mean when talking about *tempo* $T(E)$ *at a determined position* E on the score, i.e.

$$T(E) = \frac{1}{de/dE(E)}[♩/min]$$

where we have chosen the units quarter notes (♩) and minutes (min), as usual in Western music.

It is astonishing that the common understanding of tempo in music is still somewhat akin to what was the state of the art in physics in Galileo's

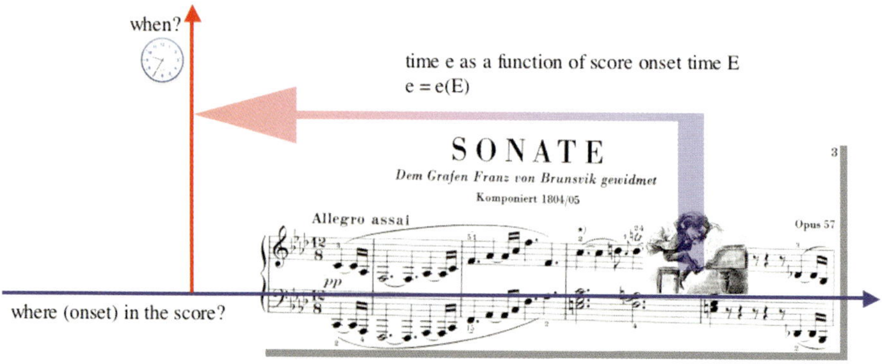

Fig. 6.4. When a performer moves through a score, physical time e can be seen as a function of the score's symbolic time E, the momentous position of the performing artist.

times. This means that tempo cannot be conceived as a momentous slope of a continuous curve, but must be thought as being locally constant. In other words, tempo is a step function that changes its values in discrete times. This view is made evident in Hermann Gottschewski's graphics [21] used to describe tempo variations for piano roll recordings (figure 6.5).

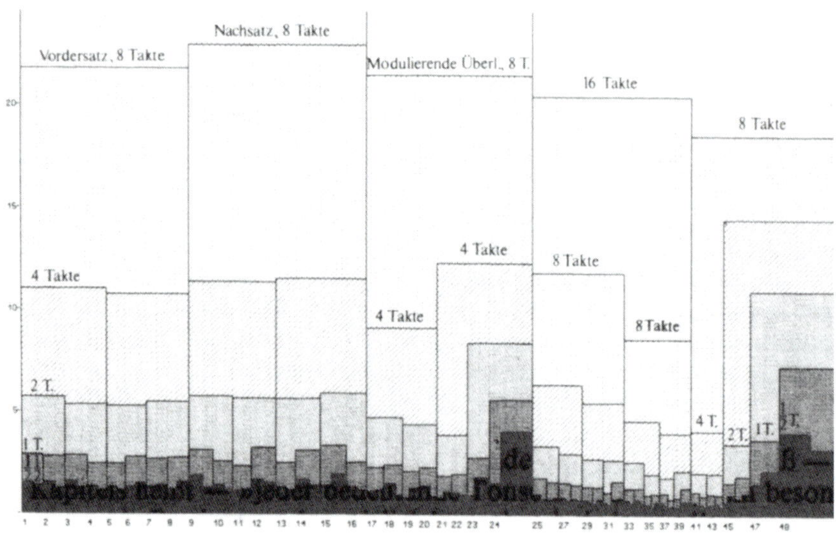

Fig. 6.5. Hermann Gottschewski's locally constant tempo curves.

In the ongoing experimental research on tempo performance, as investigated by Alf Gabrielsson [41], for example, tempo representation has already moved from a step-wise representation to continuous—more precisely to the polygonal representation as a concatenation of linear curves of inverse tempo (figure 6.6). Here the average tempo is shown as level 0 percent for a performance of Mozart's piano sonata KV 331, A major. The local deviation of tempo (every onset of a melody note being measured) is indicated by percentage in terms of duration, i.e., the inverse of tempo. The linearly interpolated curve therefore relates to $1/T$, what means that the tempo is interpolated by $\frac{1}{aE+b}$ between the measured points. The result of this measurements shows two curves, one solid, the other dotted. Both relate to performances of that piece by one and the same performer. They show that tempo is shaped in a systematic way, since the curves are very similar to each other, resulting in a typical rallentando (i.e., increasing note durations) at the end of the first and second phrase of the period.

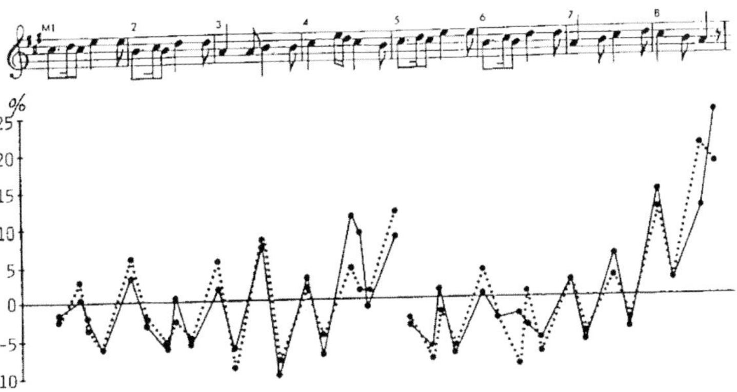

Fig. 6.6. Two $1/T$ curves from performances of Mozart's piano sonata KV 331, A major, by the same performer.

In musical practice and notation, tempo has been denoted in three different ways:

- by absolute tempo signs, such as Maelzel's metronomic notation—e.g. "M.M. ♩ = 120", meaning that tempo is 120 ♩-s per minute—or else verbal descriptions such as *andante*, *adagio*, etc.
- relative punctual tempo signs such as *fermata*, *general pause* (G.P.), *caesura*, *breath*, *mark*, etc.
- relative local tempo signs such as (1) coarse indications like *ritardando*, *rallentando*, *accelerando*, *stringendo*, etc., (2) notation of correspondence between adjacent tempi, such as, for example, 𝅗𝅥 = ♩., which means that M.M. ♩ = 93 is replaced by M.M.* ♩ = $93\times(1/2)/(3/8) = 93\times 4/3 = 124$, (3) signs of type *a tempo* or *istesso tempo*, re-establishing the original tempo.

6.1.1 Some Remarks on the History and Ethnography of Tempo

In the Western musical tradition, the practice of notating tempo has evolved slowly since approximately 1600 [122]. As pitch and rhythmic notation developed and became more precise, attempts were made to do the same with tempo indications. However, as musicians and conductors honed their skills to interpret the symbols of notated rhythm and pitch as accurately as possible, any attempt by a composer to exercise the same control over the tempo of a composition was (and is, to this day) met with much resistance. This may be (in part) due to the fact that given all of the factors of a piece that can be manipulated by a performer, tempo is the most malleable.

The first attempt to codify tempo markings happened between 1600 and 1750 during the Renaissance in Italy [31]. These descriptive words (*presto*, *andante*, etc.) are still used today as general indicators of tempo. However, the definitions of these words are highly subjective; attempts to clarify exactly what they mean (i.e. with a metronome marking) seem to go unnoticed by those who are performing the music to which they are attached. This confusion is partly due to the fact that the range of these terms implies a 1:8 ratio. That is, the fastest tempo (*prestissimo*) will be eight times faster than the slowest (*molto-adagio*) [119] . Curt Sachs in [119] also points out that the physical tempo of the piece (the tempo at which a conductor will beat quarter notes in 4/4) is a much smaller ratio than 1:8; it is more like 1:4. This is not to say that a piece cannot be thought of as having a tempo outside that range, and it does stand to reason that on the extreme ends of the range (i.e. less than M.M. 50 b.p.m. (beats per minute) and more than 140 b.p.m.) both performer and conductor alike will either "half" or "double up" their physical beat depending on which direction the tempo changes.

One would think that the metronome (invented by Johann Nepomuk Maelzel and made commercially available in 1815 [99]), would alleviate the uncertainty that previously existed in regard to "correct" tempi. Now that composers had a tool to mark exactly how fast or slow they wanted a particular composition, there could be no more of the ambiguity that allowed for such wide interpretation by performers. This has proven not to be the case, however. Some composers, such as Brahms, refused to use Maelzel markings because they were too rigid. And others, such as Beethoven, put different markings on different copies of the same score [122]. This inconsistency from composers, combined with doubts of the early metronomes' accuracy and the tendency of publishers to make adjustments to Maelzel markings according to their personal taste, have made it difficult to trust (or provide a justification to ignore) the metronome markings on scores from the early days of the metronome.

Even in modern scores, Maelzel markings are still understood to be a guideline more than a rule when it comes to tempo. Other factors, such as the performer's ability, artistic intuition, and personal taste, are seen as more important in tempo determination than the composer's notion of the ideal tempo. One cannot help but question this attitude that is prevalent among music practitioners. There is evidence that listeners are able to detect a slowing of tempo more readily than a raise in tempo and that the natural preference of both listeners and performers alike is toward faster tempi [147]. From this we can gather that we are naturally more comfortable and more in tune with fast tempi. This phenomenon seems to also be illustrated in the activities of students who are learning a new exercise or piece of music. Their tendency is to go faster and faster. For some reason they equate speed with success.

We must also take into account historical aspects of tempo that do not fall under the purview of "European Art Music." For example, in jazz there exists a similar ambiguity as to the notation of tempo. There are some conventions similar to those of the Italian descriptors used in the European tradition. However, the terms are less codified and equally vague. For instance, "Medium Swing" could refer to anything played from M.M. 100 - 188. Other terms such as "Medium Up," "Up," and "Ballad Tempo" are also used. They too are approximate in their meaning. In addition, some composers take it upon themselves to create new tempo designations like, "quasi-walking-latin-ballad," or "tempo di A-train." The latter refers to a song (*Take the A-Train* by Duke Ellington) that has been recorded hundreds of times at hundreds of different tempi, and the former seems to be an attempt to define tempo in the spirit of Beethoven's *C Major Mass* op.86, which was inscribed "Andante con moto assai vivace quasi Allegretto ma non troppo." In both cases, the descriptive markings seem to do more to cloud the issue than to clarify it.

Another non-European terrain that deserves exploration is that of the music of Africa. As Westerners, we cannot purport to have a clear understanding of the subject based solely on musicological readings. And research on the integration of African concepts of tempo and time with those of the Western tradition is limited, to say the least. Most of the writing on tempo to this point has been in regard to its mechanical function in Classical music (i.e. what bearing it has on the musical experience as a whole) or semantic matters (i.e. how to determine the "proper" tempo). And most of the writing on African concepts of time and tempo regards how it functions within itself. The African concept of tempo and time, in general, is much different—and as we understand it—is much closer in spirit to something that resembles David Epstein's [29] definition of tempo[1], in that it is inextricably linked to the everyday lives of all of the music's participants. It is fully integrated into the existence of the

[1] "Tempo is yet more complex as a phenomenon, for it embodies more than pacing per se. Tempo has generally been acknowledged as a consequence of the sum of all factors within a piece—the overall sense of a work's themes, rhythms, articulations, 'breathing,' motion, harmonic progressions, tonal movement, [and] contrapuntal activity. In this respect tempo is a product of the music—its Gestalt, so to speak, as well as an element of that Gestalt. Yet tempo, in its capacity as pacing, is another product still—a reduction of this complex Gestalt to the element of speed per se, a speed that allows the overall, integrated bundle of musical elements to flow with a rightful sense."

participants and is an inextricable part of the whole of their being. In effect, it has taken Epstein's supposition to a higher level and applied those axioms to a grander concept of "everything" as opposed to keeping it localized to just the performance of music. The subject of adapting this philosophy and attitude to the Western tradition bears further investigation.

The history of tempo is a massive subject on which many writers have expounded with great skill and academic rigor. We have no illusions that this history is complete in any sense of the word. It is our hope, however, that we have provided an adequate cross section of some of the work that has been done on this subject to date. And that if the reader is interested in expanding his/her knowledge of the subject, we have provided an interesting starting point as well as planted seeds for new research and philosophical inquiry.

6.2 Calcuating Time from Tempo

Let us now terminate the tempo discussion with the question of how time relates to tempo. Intuitively: If I drive my car, starting at position S_0 on the street, and if I observe the speed on my speedometer until I arrive at position S_1, how much time has then elapsed? Musically, this means the following: If I play a piece, starting at onset time E_0 on the score, and if I observe the music's tempo on my tempo curve until I arrive at position E_1, how much time (in seconds, say) has then elapsed?

Let us look at figure 6.7 for this calculation. To the left, we see the physical time as a function of symbolic time, the onset position on the score. The physical time that elapses between the initial physical onset $e_0 = e(E_0)$ and the final onset $e_1 = e(E_1)$ is $\int_{E_0}^{E_1} \frac{de}{dE}(E)dE$. But $\frac{de}{dE} = \frac{1}{T}$, whence

$$e_1 - e_0 = \int_{E_0}^{E_1} \frac{1}{T(E)} dE.$$

Let us give two representative examples:

1. Constant tempo $T(E) = T = \text{const}$. Then we have

$$e_1 - e_0 = \int_{E_0}^{E_1} \frac{1}{T(E)} dE = \frac{E_1 - E_0}{T}.$$

2. Linear tempo $T(E) = T_0 + S \cdot (E - E_0)$, $S \neq 0$, where $S = \frac{T_1 - T_0}{E_1 - E_0}$ with $T_0 =$ tempo at E_0 and $T_1 =$ tempo at E_1. Then

$$e_1 - e_0 = \int_{E_0}^{E_1} \frac{1}{T_0 + S \cdot (E - E_0)} dE = \frac{1}{S} \cdot \ln(\frac{T_1}{T_0}).$$

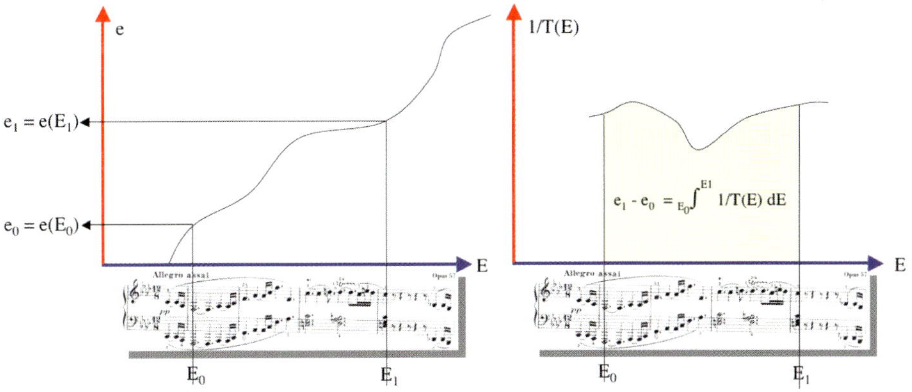

Fig. 6.7. Calculating time differences from inverse tempo curves.

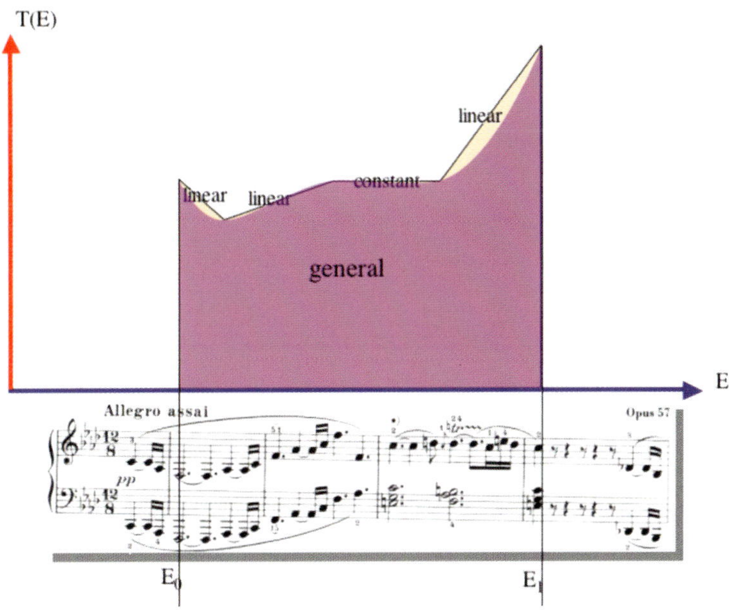

Fig. 6.8. Polygonal approximation of general tempo curves by linear and constant tempi.

With these results we may calculate the general case by sufficiently fine polygonal approximation of the tempo curve, as in figure 6.8.

We have not said the last word about the concept of tempo—we shall come back to this later—but want to emphasize that there are a number of open questions, such as:

- Does every piece have a tempo?
- Could it happen that a piece has several simultaneous tempi?

- Could it be that different parameters have a specific relations to their unfolding in time, and thereby enforcing a multplicity of tempi?
- Are all tempi, if there are several such things, equivalent, or may it be that we have herarchies of tempi, such as the conductor's tempo versus the musicians' or the soloists' tempi?

7

Tuning, Intonation, and Dynamics

Opportunity is often missed
because we are broadcasting
when we should be tuning in.

Tuning and intonation deal with the transformation of pitch symbols to frequency, while dynamics does so with the symbols of loudness that are transformed to sound pressure. Mathematically speaking, both transformations behave similarly to the situation with tempo. However, the conditions for tuning and intonation are much more complex than for tempo with regard to the semantics of pitch symbols. The dynamical situation is by far more simple, and we therefore have included this topic in the present chapter.

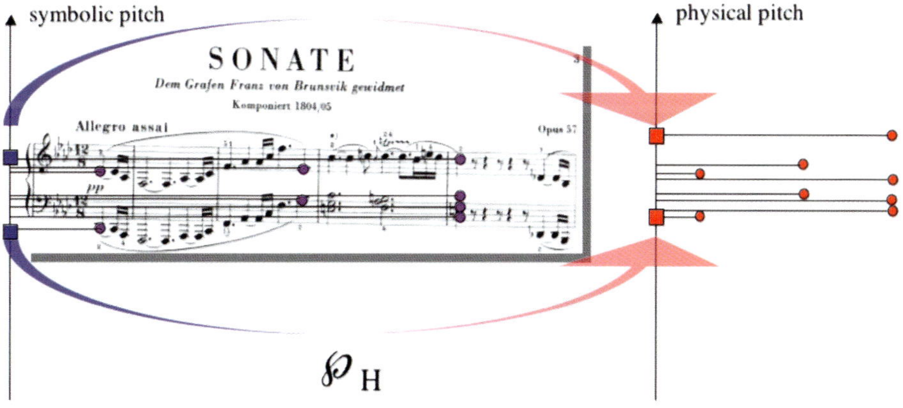

Fig. 7.1. Tuning and intonation deal with the transformation of pitch symbols to physical pitch, or, equivalently, frequency.

G. Mazzola, *Musical Performance*, Computational Music Science,
DOI 10.1007/978-3-642-11838-8_7, © Springer-Verlag Berlin Heidelberg 2011

7.1 Tuning and Intonation

Tuning and intonation (see figure 7.1) define the performance of pitch in two ways: lobally and locally tuning deals with the global background of pitch calibration or gauging. This means that we have to specify the frequencies associated with pitch symbols as a general setup *before* playing the concrete notes. For example, a piano performance would use the piano's tuning (usually set up by a professional technician) that remains fixed throughout the entire performance (except if the pianist detunes the instrument by brute force or by compositional directives). Given the relatively stable tuning background, intonation is the shaping of frequencies that is performed in local situations during the playing of the piece and happens to affect frequency deformations for expressive purposes. Although intonation is usually small in quantity, measurements of singers' intonation (not the vibrato, but the determination of pitch) have demonstrated dramatically large deformations, sometimes up to a half tone (!) [11].

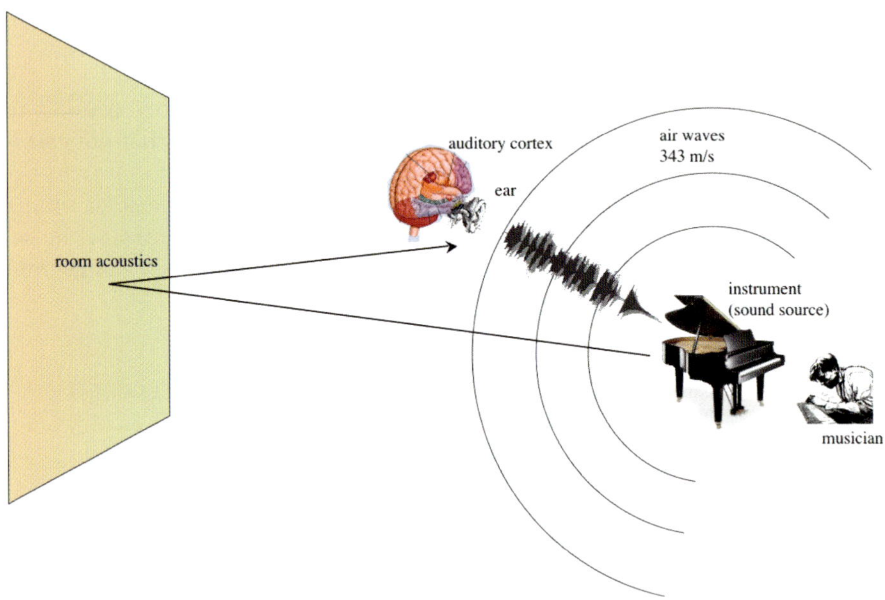

Fig. 7.2. Sound travels through a complex pathway before it reaches the human ear and is processed in the auditory cortex.

In order to understand pitch performance, we first have to look at the space of pitch, which is much more complicated than naive thoughts would make us believe. To begin with, the instrumental sound production in music needs an instrument, a musican who interacts with the instrumental "interface," and a room filled with air where the sound wave expands at a normal speed of 343 m/sec from the instrumental source, reaching a number of walls (ceiling, floor, etc.), from where the sound is reflected in a more or less distorted and delayed way (figure 7.2). The original sound and its reflected second variants add up to the sound that penetrates the human ear, reaches the eardrum, is reinforced by a factor of 20 by the mechanical arrangement of three tiny ossicles (mallesu, incus, stapes), is then transferred to the cochlear spiral of the inner ear, which is filled with lymphic liquid and transfers the liquid's wave to the Corti organ that finally incites the auditory nerve, which leads the excitation to the auditory cortex in the Heschl gyri of the temporal lobe (see [84] for details).

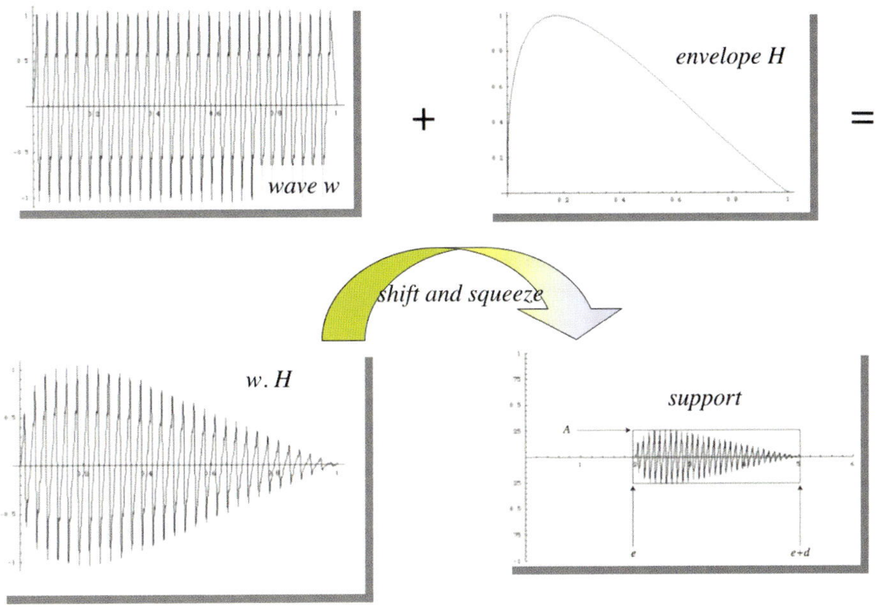

Fig. 7.3. The concrete sound event is built from an infinitely extended wave, an envelope function and its shifted and squeezed deformation defining the actual onset time e, duration d, and amplitude A.

These sound waves, when measured in a determined place, such as the eardrum, can express what humans perceive as pitch. In a standard but quite simplistic model of sound, this perception relates to the frequency of a sound burst. The sound unit is described as follows (see figure 7.3): We first consider an infinitely lasting variation $w(t)$ of air pressure, which is periodic in time t, i.e., there is a period P time such that $w(t) = w(t + P)$ holds identically. The inverse number $f = 1/P$ is called the wave's frequency, its unit being Hertz—$Hz = 1/sec$, the inverse of the time unit sec, the second. The finite extension of a realistic sound is achieved by the multiplication of the wave with an envelope function $H(t)$, a positive function with maximal value $= 1$ that vanishes outside the time unit interval $I = [0,1]$ of times $0 \leq t \leq 1$. The sound package is then shifted and squeezed to onset time e, duration d, and amplitude A (the maximal displacement from normal air pressure). The formula for the resulting sound is

$$p(t) = A.w(t).H((t-e)/d).$$

In musical acoustics (see [84, Part XV] for details), frequency and amplitude are not used directly but mediated by logarithmic functions, defining pitch and loudness, respectively. The logarithm is used because humans perceive equal ratios of frequencies or of amplitudes as being equal and not their differences, this is the famous *Weber-Fechner law* of psychoacoustics. Logarithm simply turns such ratios into differences and therefore transforms physical units to units that are more meaningful to the human ear's perception. More precisely, we have these transformations:

$$h = pitch(f) = \frac{1200}{\log_{10}(2)} \log(f) + const. \text{ [Ct] Cent}$$

$$l = loudness(A) = 20 \log_{10}(\frac{A}{A_0}) + const. \text{ [dB] deciBel}$$

$$A_0 = 2 \cdot 10^{-5} \; [N/m^2] \text{ (hearing threshold pressure).}$$

The reason for that strange factor $\frac{1200}{\log_{10}(2)}$ is that when we increase by one octave, which is to say we double the frequency, then the pitch differs by 1200 [Ct] (Cent), meaning that an octave is divided into 1200 Cent steps. And 100 of them add up to one of 12 semi-tone steps within an octave in what is defined as *12-tempered tuning*. It is in fact a tuning, since the semi-tone steps on the symbolic score level are mapped to physical pitch steps of 100 Ct each. The frequency ratio for such a semi-tone step is $\sqrt[12]{2} \approx 1.05946....$ In other words, the *12-tempered tuning* attributes to the sequence of symbolic semi-tone steps the sequence of pitches (in Ct) and corresponding frequencies (in Hz) of shape $f(o) = f_0 \cdot (\sqrt[12]{2})^o = f_0 \cdot 2^{o/12}$, where $o \in \mathbb{Z} = \{\ldots -3, -2, -1, 0, 1, 2, 3 \ldots\}$ is an integer and f_0 is a reference pitch defined by that constant. In music technology, the MIDI standard selects the middle C of the piano as corresponding to the integer $o = 60$, with other values reaching from $o = 0$ to $o = 127$. The constants for pitch and loudness depend upon the technical setup defining reference pitch or loudness and are not important except for gauging purposes. We shall come back to this later.

Besides the 12-tempered tuning, music theory and practice has known a huge number of tunings, which express theoretical and/or historical perspectives. We will not delve into this topic as such, because it is not a performance theme, but prefer to elaborate on a deeper problem relating to the spaces involved in tuning theories and the question of how to perform in such contexts. The prototypical non-tempered tuning is the classical just tuning. It stems from the ancient Pythagorean approach that is motivated by the myth of the

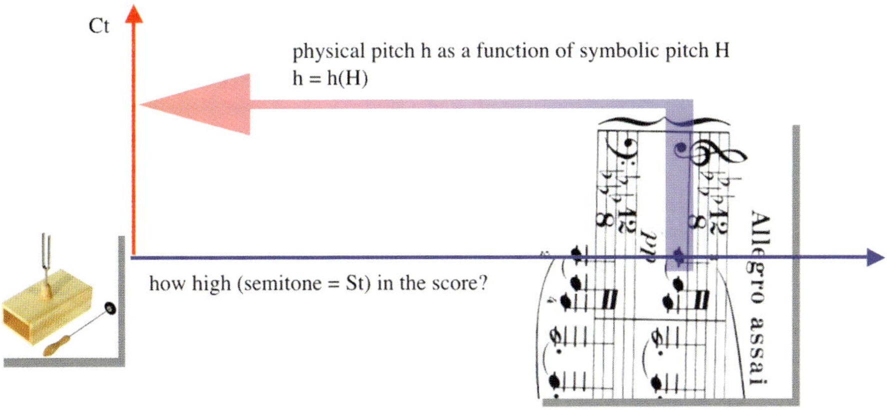

Fig. 7.4. Intonation describes the locally variable transformation of pitch symbols measured in semi-tone units (St), into physical pitch values, measured in Cents (Ct). The reference of such transformation is usually a (mechanical or electronic) tuning fork, a mechanical sample is shown to the left of the image.

transcendental tetractys. This symbolic triangle is built from ten points (ten being a holy number in ancient Greece) that expresses frequency ratios of consonant intervals by small numbers. The triangle has four layers: The basis is a row of four points, upon which three points are layered, then two, and on top there is a single point. The successive numbers 2:1, 3:2, 4:3 represent the frequency ratios of the octave, fifth, and fourth. Just tuning takes up this data, includes the major third ratio 5:4 added to the Pythagorean system at the end of the Middle Ages, and is defined by a general formula of frequencies using those first three prime numbers 2, 3, and 5, and a reference frequency f_0:

$$f(o, q, t) = f_0 \cdot 2^o 3^q 5^t, o, q, t \in \mathbb{Z}$$

Here is the table for the frequency ratios of 12-tempered and just tuning starting from tone c. For example, the frequency $f_0 \cdot 2^{-5}3^2 5^1$ represents the tone $f_\sharp$ in just tuning, i.e., the frequency ratio $45/32$ of the tritone with respect to tone c.

Tone name	Frequency ratio	Octave coord.	Fifth coord.	Third coord.	Pitch (Ct)	% deviation
c	1	0	0	0	0	0
$d_\flat$	16/15	4	-1	-1	111.73	+11.73
d	9/8	-3	2	0	203.91	+1.96
$e_\flat$	6/5	1	1	-1	315.65	+5.22
e	5/4	-2	0	1	386.31	-3.42
f	4/3	2	-1	0	498.05	-0.39
$f_\sharp$	45/32	-5	2	1	590.22	-1.63
g	3/2	-1	1	0	701.96	+0.28
$a_\flat$	8/5	3	0	-1	813.69	+1.71
a	5/3	0	-1	1	884.36	-1.74
$b_\flat$	16/9	4	-2	0	996.09	-0.39
b	15/8	-3	1	1	1088.27	-1.07

Before we go into those difficult questions about performing different tunings, let us elaborate the analogy of tuning/intonation with tempo. We do intentionally include intonation, too, not only fixed tuning. In analogy to tempo, this means that we are also looking for the locally variable mapping from the score's pitch data to the physical pitch. To begin with, the transformation of symbolic pitch (measured in semi-tone units) to physical pitch (measured in Cents) is shown in figure 7.4.

This map is a priori defined for the whole continuum of symbolic pitch, not just for the discrete semi-tone values of a classical score. There are good reasons for this (as there were reasons for defining the performance of time for all real-valued times). To begin with, microtonal intervals are used in many compositional contexts, be it in different ethnic or compositional styles. Moreover, glissandi have been composed in completely classical contexts. So the entire real interval between two limiting notes of a glissando are in fact symbolic pitch values that need to be performed into intervals of sound pitch. Again, this transformation is supposed to have a slope in each point, and we get this image, figure 7.5.

In this setup, completely analogous to the tempo setup discussed in section 6.1, where h stands for physical pitch and H for symbolic pitch, speed takes the form

$$speed(H) = \frac{1}{dh/dH(H)}$$

We move through the symbolic pitch space as a function of physical pitch. This is precisely what musicians mean when talking about *intonation* $S(H)$ (S for German "Stimmung") at a determined position H on the score, i.e.

$$S(H) = \frac{1}{dh/dH(H)} \text{ [St/Ct]}.$$

A first situation of this formalism is the 12-tempered tuning, as shown in figure 7.6. Here we have a reference pitch given by the concert pitch for 440 [Hz], corresponding to symbolic pitch a', the number 69 in MIDI notation.

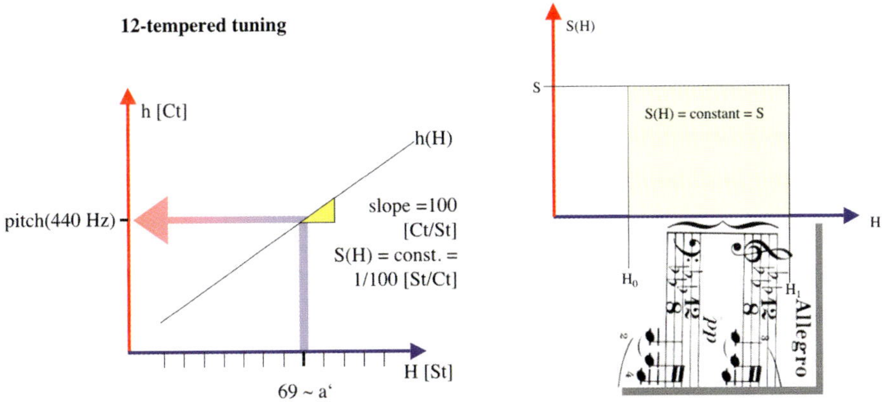

Fig. 7.6. The 12-tempered tuning curve is a linear function with slope 100 [Ct/St]. The intonation function is therefore $S(H) = \text{const.} = 1/100$ [St/Ct].

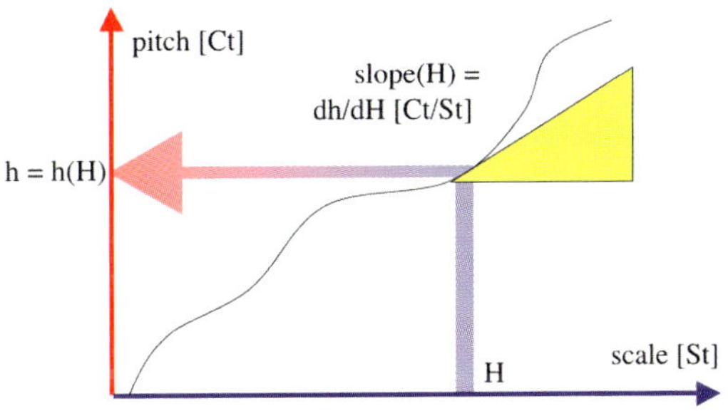

Fig. 7.5. Speed is viewed as the inverse slope of the curve defined by physical pitch being a function of symbolic pitch.

The situation for just tuning (and related intonation) is more delicate. Here we have a slight deviation from the 12-tempered curve (figure 7.7). But this representation as a map from the real line of symbolic pitch to the real line of physical pitch is not what corresponds to intrinsic musical thinking. The latter does not view symbolic pitch as living in a one-dimensional real line space. But how is this possible when all symbolic pitches somehow seem to be representations of (logarithms of) frequencies, which live in a one-dimensional real line space?

The point is that in musical thinking, the octave, the fifth, and the major third are conceived as independent "directions" in the harmonic space of pitch relations. This approach is not only a wishful illusion of music theorists; it has been realized by the theory of pitch developed by the great mathematician Leonhard Euler in the eighteenth century in his work with the speaking title

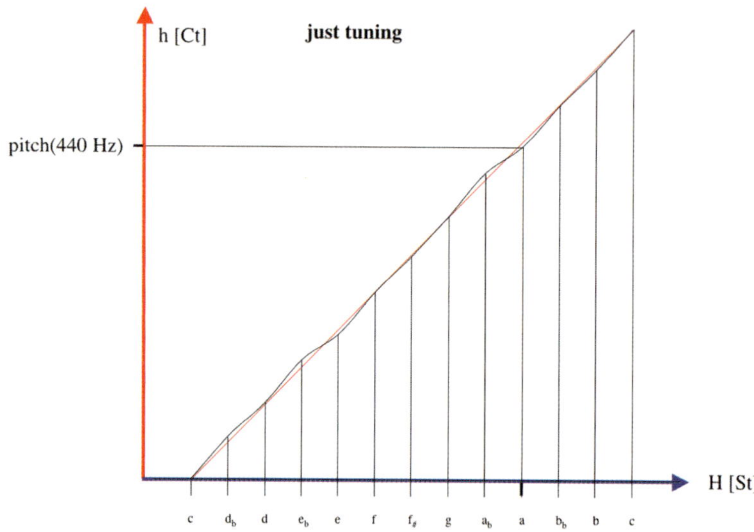

Fig. 7.7. The just tuning curve is a delicate deformation of the 12-tempered one. Its representation on the real line spaces of pitch is not what musical thought is about. A more adequate space would be Euler's three-dimensional space. But this poses difficult problems for performance.

speculum musicae [30], meaning the visualization of music. In this theory, Euler attributes to octave, fifth, and major third three independent vectors three-space.

This is possible for the following reason. Take the above formula defining just tuning: $f = f(o, q, t) = f_0 \cdot 2^o 3^q 5^t$. If we pass to the logarithm of frequency representing pitch, we have $\log(f) \sim o \cdot \log(2) + q \cdot \log(3) + t \cdot \log(5)$, a linear combination of three "vectors"—$\log(2), \log(3)$, and $\log(5)$. It can be shown that these three numbers behave like spatial vectors[1] as long as the coefficients o, q, t are rational numbers, i.e. fractions of integers such as $3/4$. But this is what we have in just tuning—we even have integer coefficients. And recall that 12-tempered tuning is also included in this setup: just take $q = t = 0$ and run through all rational octave coefficients of the form $o = x/12, x \in \mathbb{Z}$.

The distribution of the twelve notes of a chromatic just scale is shown in figure 7.8. Although they appear to be distributed quite wildly, it can be shown that they share a unique symmetry that is even related to contrapuntal theory [84, Chapter 30]. In this representation, a 12-tempered chromatic scale would be built from 12 equidistant points on the unit interval in octave direction, starting from c.

The problem of performing in Euler's space is that we are representing symbolic pitch as being points in three-space, which requires more than just

[1] They are linearly independent, i.e. $o \cdot \log(2) + q \cdot \log(3) + t \cdot \log(5) = 0$ holds if and only if $o = q = t = 0$.

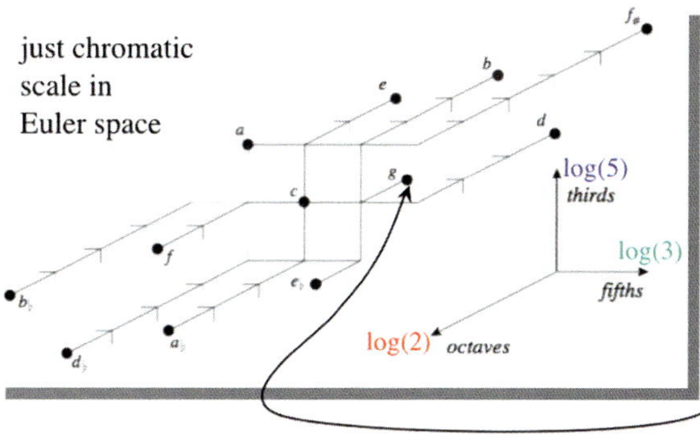

Tone name	Frequency ratio	Octave coord.	Fifth coord.	Third coord.
c	1	0	0	0
d♭	16/15	4	-1	-1
d	9/8	-3	2	0
e♭	6/5	1	1	-1
e	5/4	-2	0	1
f	4/3	2	-1	0
f♯	45/32	-5	2	1
g	3/2	-1	1	0
a♭	8/5	3	0	-1
a	5/3	0	-1	1
b♭	16/9	4	-2	0
b	15/8	-3	1	1

Fig. 7.8. In Euler's space that is spanned by the logarithm "vectors" of the octave (2 for 2/1), fifth (3 for 3/2), and major third (5 for 5/4), the twelve pitches of the chromatic scale appear as a spatial arrangement of points.

the frequency-related real number. In fact, we would need three real coordinate numbers, not only one, to play such pitch points. This problem will be addressed in chapter 19 on string quartet theory. It deals with the introduction of additional instrumental parameters beyond onset, pitch, loudness, and duration. For the time being, we will stick to the simpler real line model of pitch.

Because tuning and intonation has many formal parallels with tempo, a number of open questions can be raised:

1. Does every piece for a single instrument have a tuning/intonation?
2. We may have different coexisting tunings for orchestral music. For example when piano and violin are collaboration, we will have the 12-tempered versus a variable, somehow just tuning. How would these be modeled?
3. Are there hierarchies of tuning? Especially in large orchestral works?
4. What about the strucure (and function) of variable tuning and intonation in non-European music, such as Egyptian Maqam music?
5. What happens with tuning for instruments with variable tuning when we modulate tonality?

Some of them, but not all, will be discussed and solved in the next chapter. We should listen to a number of examples (see our example list on page 263) for different tunings, such as

A. Schumann's Kinderlied op.15/1 and Webern's op.27/II in these tunings:

- Pythagoren (no 5 component)
  Schumann's Kinderlied op.15/1 ♫ 2
  Webern's op.27/II ♫ 3
- meantone (following Pietro Aron)
  Schumann's Kinderlied op.15/1 ♫ 4
  Webern's op.27/II ♫ 5

- well-tempered (following Bach)
  Schumann's Kinderlied op.15/1 ♫ 6
  Webern's op.27/II ♫ 7
- slendro (Balinesian)
  Schumann's Kinderlied op.15/1 ♫ 6
  Webern's op.27/II ♫ 7
- 12-tempered
  Schumann's Kinderlied op.15/1 ♫ 8
  Webern's op. 27/II ♫ 9

B. Example of Egyptian improvised Maqam music (recording ♫ 12 and transcription by James Holdman), with microtonal alteration signs (half flat, for example, in the buzuq voice).

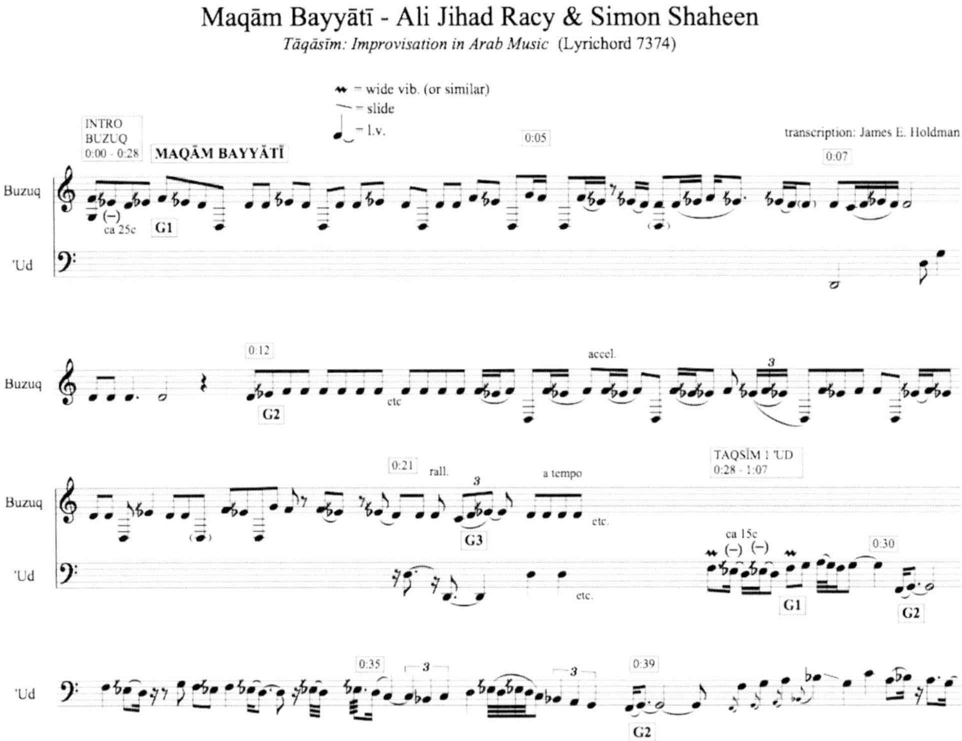

Fig. 7.9. Transcriptions of Egyptian Maqam music by James Holdman.

C. Microtonal music: Alois Hába (1893-1973): quarter- and fifth-tone compositions for strings ♫ 13.

7.2 Dynamics

The hearing threshold pressure A_0 is that minimal (relative) pressure level where we hear a just noticeable sound. In Western notation, the loudness quantity (in dB) is associated with a small number of dynamical symbols, which corresponds to the tuning transformation in pitch performance. Usually, these symbols range from five-fold pianissimo and five-fold fortissimo, yielding this list of increasing dynamical symbols:

ppppp, mpppp, pppp, mppp, ppp, mpp, pp, mp, p, mf, f, mff, ff, mfff, fff, mffff, ffff, mfffff, fffff

Although there is no precise convention about the loudness steps between the physical loudness associated with these symbols, one may suppose that these steps (in dB) are all equal. In the MIDI standard for dynamics, the MIDI symbols (called *velocity*, because they remind us of the velocity of the finger hitting a key to define its loudness) reach from 0 to 127 (same as for pitch), and one would then have an equidistant distribution of MIDI values to represent the above list. This means that we could start with *ppppp* ~ 1 (0 for silence) and increase by steps of 7 MIDI velocities, yielding *mpppp* ~ 8, *pppp* ~ 15, and so on through *fffff* ~ 127. Observe that while neither MIDI pitch nor MIDI velocity are physical quantities, they still need a gauging on the MIDI-capable expander to signify physical sound attributes. Figure 7.10 gives a number of physical loudness values associated with musical loudness. We should address also the frequent argument that the dynamical symbols in musical notation are heavily dependent on the context, and that it is therefore wrong to attribute a well-defined numerical value to them. This is true, but it is exactly the difference between nominal and performed dynamical values that defines the values that are context-dependent. The context is the analytical, emotional, and gestural rationales shaping dynamics in performance. It defines the performed dynamical values. These are variations of the nominal values, performative deformations of the mechanical setup defined by the score.

Again, like with tempo and tuning/intonation, dynamics is a transformation between two real lines. On the symbolic side, dynamical values can cover entire real intervals when we deal with crescendi or when we have electronic devices that enable continuous (symbolic) dynamical ranges. Symbolic dynamics is best given in units from MIDI, which we call Vl (velocity), while physical dynamics (loudness) is measured by dB. And again, the transformation $\wp_L : L \mapsto l(L)$ of symbolic loudness L into physical loudness l is supposed to be a curve that has a slope $dl/dL(L)$ in every argument L, and we define intensity $I(L)$ [dB/Vl] at the dynamics value L as being the inverse of the slope of $\wp_L$, i.e.

$$I(L) = \frac{1}{dl/dL(L)} \text{ [dB/Vl]}.$$

Summarizing, we have these three formulas for tempo, intonation, and dynamics, which calculate the physical differences of parameter values according

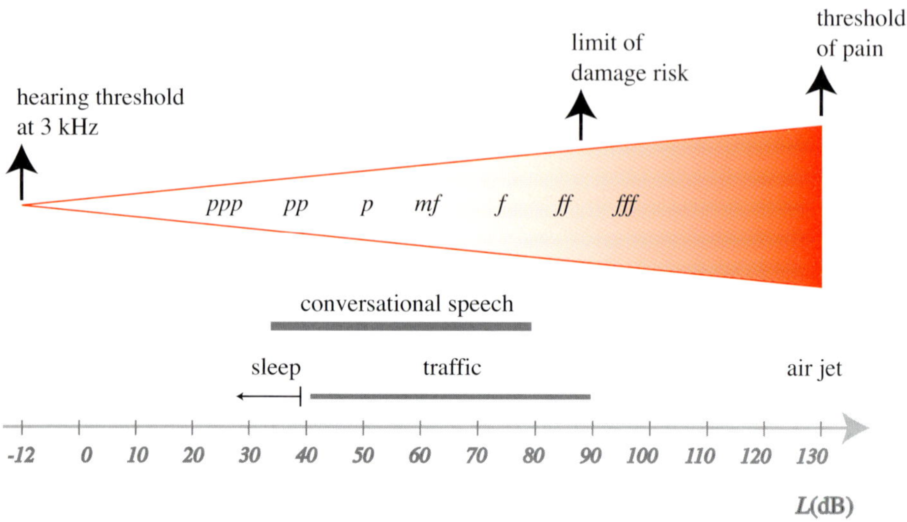

Fig. 7.10. Some physical loudness values associated with musical symbols for dynamics.

to the start and end points on the score and the given speed-related functions of tempo, intonation, and intensity:

$$e_1 - e_0 = \int_{E_0}^{E_1} \frac{1}{T(E)} dE,$$

$$h_1 - h_0 = \int_{H_0}^{H_1} \frac{1}{S(H)} dH,$$

$$l_1 - l_0 = \int_{L_0}^{L_1} \frac{1}{I(L)} dL.$$

8

Combining Tempo, Tuning, and Dynamics

A taste for simplicity cannot last for long.
Eugene Delacroix

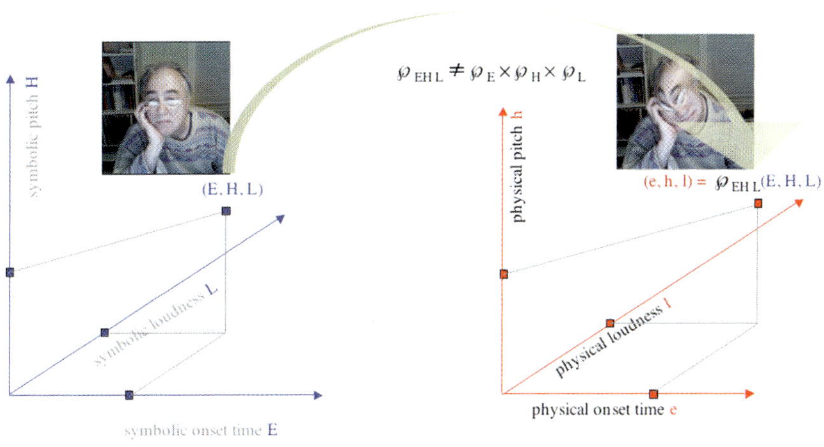

Fig. 8.1. Putting onset, pitch, and loudness together, we have a three-dimensional space map. In most practical cases, however, the map will not be a Cartesian product $\wp_E \times \wp_H \times \wp_L$ of three one-dimensional maps $\wp_E$, $\wp_H$, and $\wp_L$.

After having introduced the three basic performance maps for onset, pitch, and loudness, we have to put these components together and ask for a better understanding of the combined situation. This means that we have to look for the simultaneous performance of all these parameters. To this end, we first introduce a notation for parameter spaces: Given a sequence $P_1, P_2, \ldots P_n$ of n pa-

G. Mazzola, *Musical Performance*, Computational Music Science,
DOI 10.1007/978-3-642-11838-8_8, © Springer-Verlag Berlin Heidelberg 2011

rameters, the n-dimensional real space whose vectors $X = (X_{P_1}, X_{P_2}, \ldots X_{P_n})$ denote musical events X parametrized by the real numbers X_{P_1} for type P_1, X_{P_2} for type P_2, etc. X_{P_n} for type P_n, is denoted by $\mathbb{R}^{P_1 P_2 \ldots P_n}$. For example, the space of symbolic onset E and pitch H is denoted by $\mathbb{R}^{EH}$, while the space of physical onset e and pitch h is denoted by $\mathbb{R}^{eh}$. Often, if no confusion is likely, we write P for the coordinate X_P to ease notation.

To begin with (see figure 8.1), we would expect that the performance map $\wp_{EHL} : \mathbb{R}^{EHL} \to \mathbb{R}^{ehl}$ is the Cartesian product $\wp_E \times \wp_H \times \wp_L$ of three one-dimensional maps $\wp_E$, $\wp_H$, and $\wp_L$, i.e. $\wp_{EHL}(X_E, X_H, X_L) = (\wp_E(X_E), \wp_H(X_H), \wp_L(X_L))$. This would be true if these parameters were performed independently of one another. This may happen in an idealized, but completely unrealistic model case.

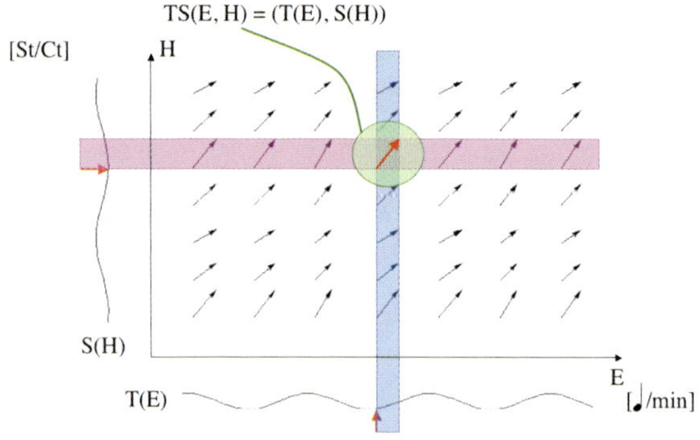

Fig. 8.2. Tempo and intonation combine to a two-dimensional vector field TS on the parameter space $\mathbb{R}^{EH}$.

Why? Because, for example, intonation is also a function of time if the instrument is not restrained to a fixed tuning, such as an unprepared piano. And the meaning of a dynamical sign is heavily dependent upon the context. For example, you may have a crescendo sign within a *mf* context. So the performed loudness of *mf* is not a constant. It is different after the crescendo when compared to the value before that sign. Also, tempo might change locally, for example in slight arpeggi or in *Chopin rubati* according to the selected voice. Chopin rubati are well-known to pianists as local variations of one voice's tempo (typically the right hand's), while the tempo of the other voice (typically the left hand's) remains constant, and the two voices meet again after that local variation. Refer to the fourth performance of a Czerny etude in example ♫ 14.

So we have to envisage more general performance mappings $\wp_{EHL}$ and we also must create a representation of such transformations that can take over the classical representation and visualization by tempo, intonation, or loudness curves as described in the previous chapters.

Let us first look at the most simple combination of two one-dimensional performance transformations: tempo and intonation. For every pair of parameter values $(X_E, X_H) \in \mathbb{R}^{EH}$, we have the two-dimensional vector

$$TS(X_E, X_H) = (T(X_E), S(X_H))$$

consisting of tempo $T(X_E)$ at onset time X_E and intonation $S(X_H)$ at pitch X_H. This defines a vector field TS on the parameter space $\mathbb{R}^{EH}$ (figure 8.2). In order to understand how to generalize this to not necessarily Cartesian product performance maps in several dimensions, we first look at the one-dimensional case of tempo (figure 8.3). Tempo at onset time E is given by a backward

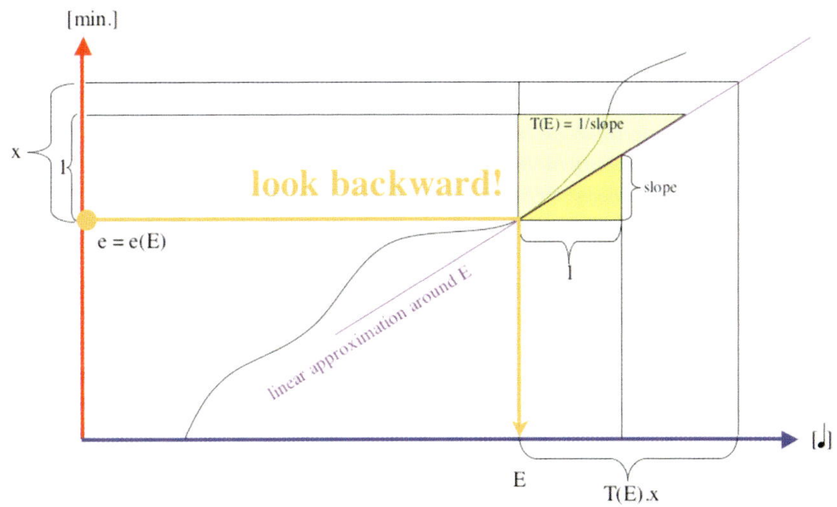

Fig. 8.3. The musical tempo relates symbolic time to physical time.

perspective on time. We first consider the linear approximation to the time performance curve $e(E)$, whose inverse slope is tempo $T(E)$. This quantity gives us the slope at e of the inverse curve $E(e)$ from physical time to symbolic time. If we take the linear approximation defined by this slope, the increase by x physical time units is linearly mapped to the increase by $T(E).x$ symbolic time units. In other words, if we move in physical time around e, that movement is linearly approximated by the movement in symbolic time that is defined by tempo. Again, this means that we move regularly in physical time (at speed 1) around e and then look at the map of this movement and its speed $T(E)$ at symbolic time $E(e)$.

With this interpretation we may easily generalize to several dimensions (figure 8.4). Instead of moving around one single dimension e, we have to move in two physical dimensions: in e and in h. This is shown with two straight curves: a horizontal one, the function $(e + x, h)$ of x at pitch level h, and a vertical one, the function $(e, h + y)$ of y at onset level e. These two straight curves are mapped backward into symbolic space $\mathbb{R}^{EH}$ and yield two curves which intersect at the point (E, H) when their parameters x, y both vanish. The tangent arrow of the backward movement with the horizontal curve is $(\partial E/\partial e, \partial H/\partial e)$ and it corresponds to the tempo quantity $T(E)$ in the one-dimensional case, while tangent arrow of the backward movement with the vertical curve is $(\partial E/\partial h, \partial H/\partial h)$. For the case discussed above of a Cartesian product $\wp_E \times \wp_H$, this gives us back the situation shown in figure 8.2. This

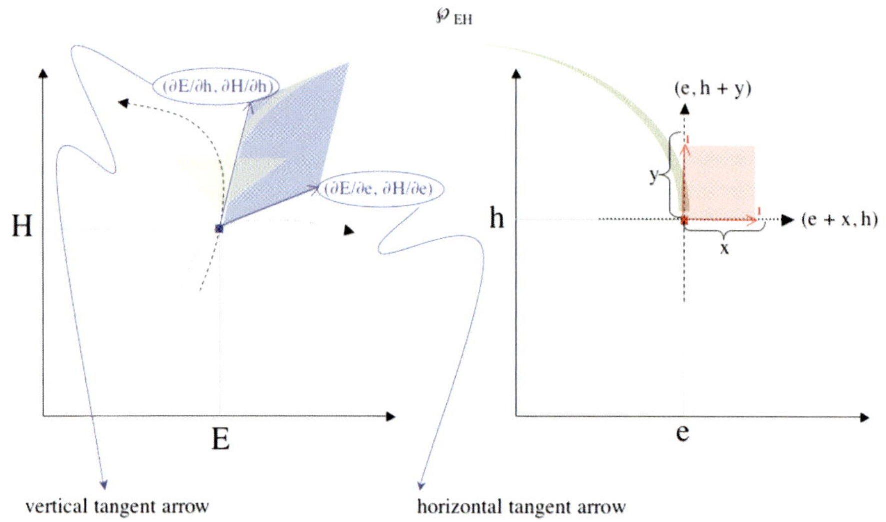

Fig. 8.4. Generalization of tempo to several dimensions: Instead of moving around one single dimension e, we have to move in two physical dimensions: in e and in h. This is shown with two straight curves, a horizontal one, the function $(e+x, h)$ of x at pitch level h, and a vertical one, the function $(e, h+y)$ of y at onset level e. These two straight curves are mapped backward into symbolic space $\mathbb{R}^{EH}$ and yield two curves which intersect at the point (E, H) when their parameters x, y both vanish.

situation is shown in figure 8.5. The horizontal curve tangent yields tempo, while the vertical one yields intonation, exactly as desired. The sum of the two tangent vectors is also the vector of the TS field previously described.

This implies that we have a perfect simulation of the two-dimensional situation derived from tempo and intonation while using only general structures, namely the horizontal and the vertical tangent vectors. But it is even better: Since the TS field is the sum of the T and the S vectors, we may reinterpret this situation as follows: take the linear map sending the horizontal unit vector $(1,0)$ to $(\partial E/\partial e, \partial H/\partial e)$, and the vertical unit vector $(0,1)$ to $(\partial E/\partial h, \partial H/\partial h)$ This is given by the so-called *Jacobian matrix* of the backward map

$$J(\wp^{-1})(e,h) = \begin{pmatrix} \partial E/\partial e \ \partial E/\partial h \\ \partial H/\partial e \ \partial H/\partial h \end{pmatrix},$$

which, when restated in terms of the symbolic variables, equals

$$J(\wp)(E,H)^{-1} = \begin{pmatrix} \partial e/\partial E \ \partial e/\partial H \\ \partial h/\partial E \ \partial h/\partial H \end{pmatrix}^{-1},$$

and the vector field $TS(E,H)$ is given by this linear map when applied to the diagonal unit vector $\Delta = (1,1)$ in $\mathbb{R}^{eh}$, namely

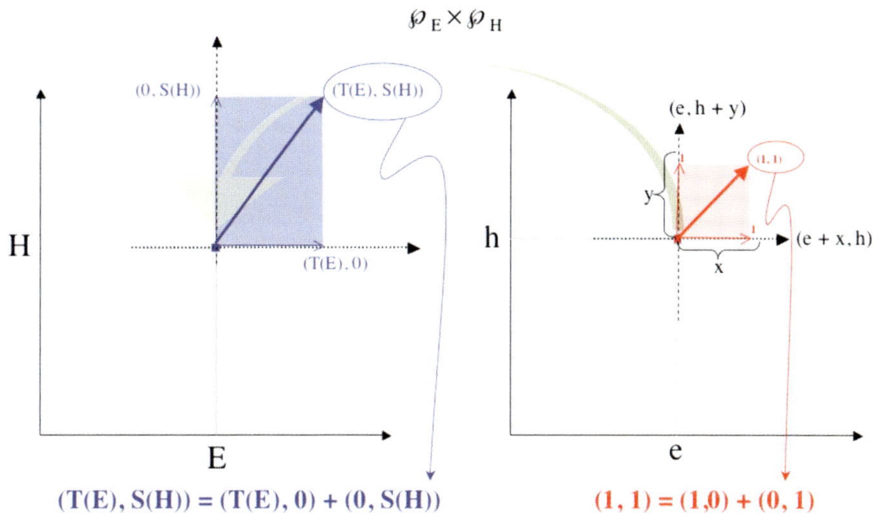

Fig. 8.5. For the case of a Cartesian product $\wp_E \times \wp_H$, our technique gives us back the situation shown in figure 8.2. This situation is shown here. The horizontal curve tangent yields tempo, while the vertical one yields intonation, exactly as desired. The sum of the two tangent vectors is also the vector of the TS field previously described.

$$TS(E, H) = J(\wp)(E, H)^{-1}\Delta.$$

This reveals the completely general method behind the construction of TS: We take the Jacobian matrix $J(\wp)(X)$ of the performance map at the point X of the given n-dimensional parameter space, then we invert it and apply this inverse matrix to the n-dimensional diagonal vector $\Delta = (1, 1, \ldots 1)$. This yields what is called the *performance field* $\mathbb{T}s$ at point X:

$$\mathbb{T}s(X) = J(\wp)(X)^{-1}\Delta$$

Figure 8.6 shows this construction. Its advantage is not only that it generalizes the local description of performance to completely generic transformations, but it also can be taken to be an *infinitesimal definition of the performance transformation*. We shall deal with this later. But for the classical situation of tempo, this is quite evident, because once the piece is being performed, the determination of physical time is completely defined from the knowledge of the tempo curve, as we have seen in the discussion of tempo in section 6.2.

For the time being, we would like to revisit the approach to performance theory forwarded by Theodor W. Adorno and Walter Benjamin [1]. Their propositions summarize as follows:

> *Walter Benjamin has defined the "power of phantasy" as being "the gift to interpolate in the infinitely small". This flashes at once the real performance.(...)*

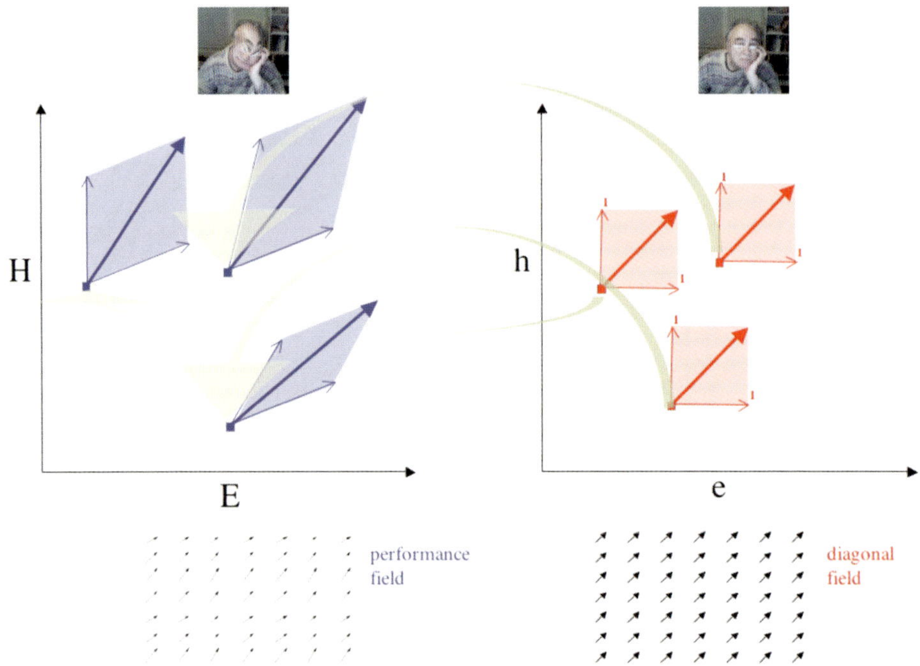

Fig. 8.6. The performance field of a performance transformation $\wp$ is the inverse image in the symbolic parameter space of the constant diagonal field with value Δ (the diagonal arrows to the right) on the physical parameter space.

> *In the densely interwoven score texture we may discover minimal cavities, where the significative performance finds it refuge.* (...)
>
> *The medium of artistic phantasy is not the decrease of precision, but the more precise.*

This text is quite arcane if one views it as a philosophical position. It is not clear what would be that "infinitely small," nor can we understand what would be the "more precise" in that perspective. Philosophy has no language and even fewer technical tools to deal with the infinitely small. The only science that can cope with such requirements is mathematics—more precisely, calculus. Calculus is the science of the infinitely small. In view of what has been conceptualized and discussed above, it turns out that Adorno and Benjamin were compeltely in the vein of calculus when asking for the infinitely small and the minimal cavities in performance. The language of performance fields perfectly meets these positions and also absorbs the classical performance fields in one dimension: tempo, intonation, and dynamics.

We shall see in the next section that the generality of the performance field theory is by no means reserved to highly sophisticated performances but arises in the most common default situation: when we consider articulation, a performance shaping that deals with duration.

9

Articulation

A bad word whispered will echo a hundred miles.
Chinese proverb

Articulation adds a completely different phenomenon to performance. While we had seen that a priori onset, pitch, and dynamics can be performed independently from each other, articulation introduces performance of duration as a situation where this new parameter is intrinsically connected to the other time parameter, namely onset. Let us first discuss performance of duration

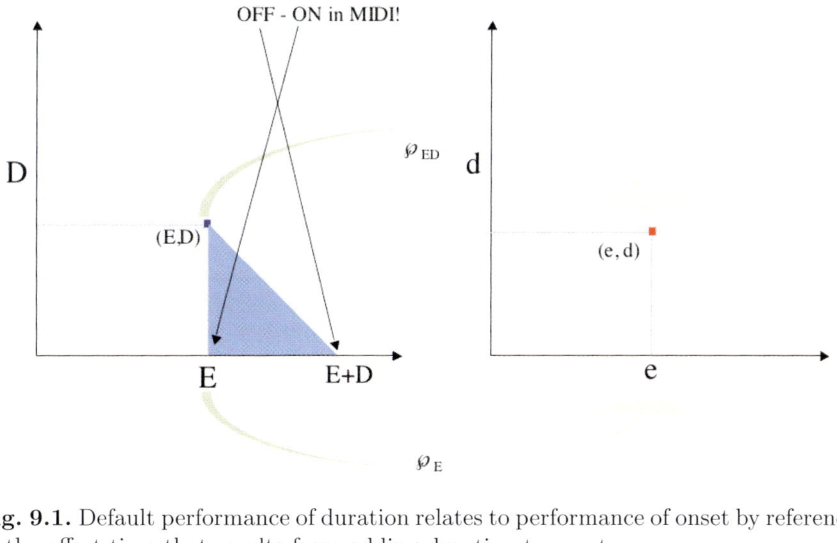

Fig. 9.1. Default performance of duration relates to performance of onset by reference to the offset time that results from adding duration to onset.

G. Mazzola, *Musical Performance*, Computational Music Science,
DOI 10.1007/978-3-642-11838-8_9, © Springer-Verlag Berlin Heidelberg 2011

without any further shaping of articulation (figure 9.1).

This situation means that the ending of a note is determined by the beginning of the next note, or pause, which means that the offset of the note with onset E and symbolic duration D is the onset $E+D$. This information implies that the physical duration d of a note with these coordinates is $d(E,D) = e(E+D) - e(E)$ (this is known as "OFF - ON" in MIDI code).

Performance of duration is therefore intrinsically connected to performance of onset. Duration is, unlike pitch, dependent on onset performance, but not vice versa—onset is a more basic parameter than duration. The default situation with onset and pitch could be described by the two projections of EH space to E and to H space

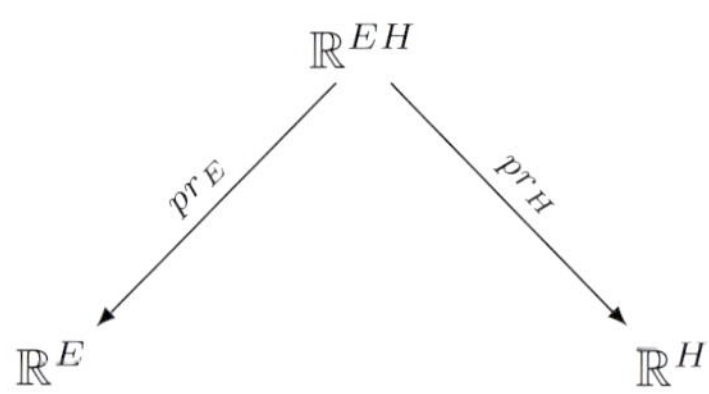

whereas the situation with onset and duration would only have one projection of ED space to E space:

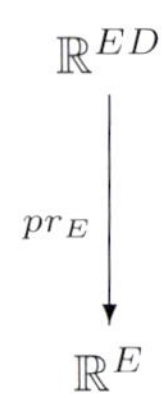

We shall make all this more precise later, but it is good to get an early idea of the hierarchy of spaces intervening in performance theory.

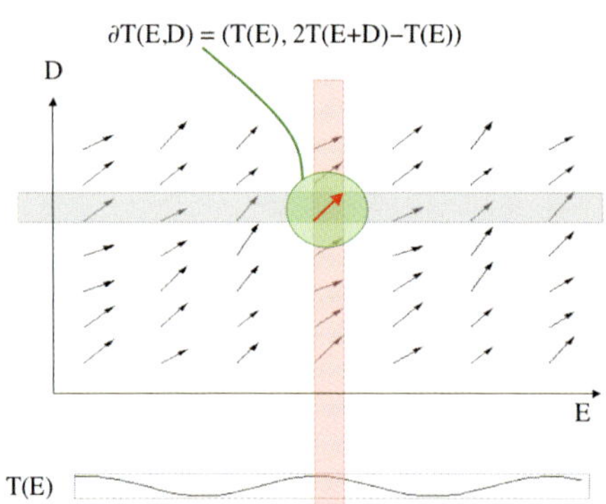

Fig. 9.2. The performance field of a default performance of onset and duration.

Now, let us calculate the performance field associated with the above performance map $\wp_{ED}(E,D) = (e(E), e(E+D) - e(E))$. The field is

$$\mathbf{Ts}(E,D) = (T(E), 2T(E+D) - T(E)),$$

and we see that its D component also depends on E, not only on D. Figure 9.2 shows this situation. The projection to onset (vertical bar) gives us the tempo field. The horizontal bar shows that the D component of the field is also a function of onset. This field is called *parallel field* and denoted by $\mathbf{Ts}(E,D) = \partial T(E,D)$ since its calculation works in parallel to the tempo curve.

So this elementary situation demonstrates that the assumption that every common field is a Cartesian product of one-dimensional fields is erroneous. The same phenomenon occurs when we calculate default performance fields for onset E and symbolic crescendo C (with associated physical crescendo c), which is a loudness change along the note much like D, but relating to L instead of E. Same with glissando G (with associated physical glissando g), which is a parallel to pitch H, and we therefore have parallel fields $\partial I(L,C)$ for crescendo and loudness, and $\partial S(H,G)$ for glissando and pitch.

To give an example of articulation, let us consider a tempo field where the default duration is multiplied by a factor $\lambda \neq 0$, i.e. $\wp_{ED}(E,D) = (e(E), \lambda(e(E+D) - e(E)))$, and such that the tempo has shape $T(E) = 1 + 0.4\sin(E)$. Then we have the articulation field

$$\begin{aligned} \mathbf{Ts}_\lambda(E,D) &= (T(E), (1+\lambda^{-1})T(E+D) - T(E) \\ &= \partial T(E,D) + (\lambda^{-1} - 1)(0, T(E+D)) \end{aligned}$$

which is the default parallel field for $\lambda = 1$, and yields a legato field for $\lambda < 1$ and a staccato field for $\lambda > 1$ (figure 9.3).

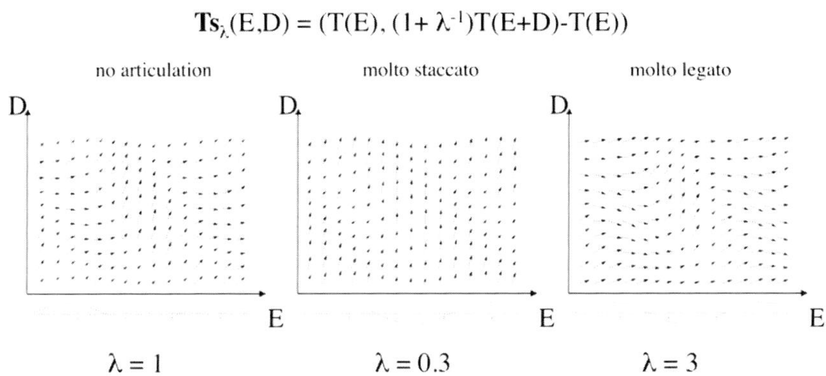

Fig. 9.3. An articulation field for a family of performances as a function of a system parameter λ.

10

General Performance Fields

La musique mathématiquement discontinue
peut donner les sensations les plus continues.
Paul Valéry [142, I]

Let us now look at the general procedure for defining performance. We write $P.$ for a sequence $P_1 P_2 \ldots P_n$. We have seen in the previous analysis that a performance $\wp_{P.}$ on a parameter space $\mathbb{R}^{P.}$ gives rise to a performance field $\mathbf{Ts}_{P.}$, and that this field can be used to define performance. We do, however, have to make this precise in the general case.

In the classical case of tempo (the one-dimensional situation of onset performance), we have the formula

$$e_1 - e_0 = \int_{E_0}^{E_1} \frac{1}{T(E)} dE$$

that yields the physical time between the initial symbolic onset E_0 and the terminal onset E_1. It completes the necessary information once we know the initial physical onset time $e_0 = \wp_E(E_0)$. So we have everything once we have that one-dimensional field $T(E)$ and the initial performance $\wp_E(E_0)$, which we denote by $\wp_E^I(E_0)$ since it is given a priori and independently from the tempo field data.

10.1 General Performance Fields

The general case works similarly but is more demanding, mathematically and musically speaking. The easiest way to understand the process is to view the performance field $\mathbf{Ts}$ as being the inverse image under $\wp_{P.}$ of the constant diagonal field Δ. This means that integral curves of the Δ field are mapped to

G. Mazzola, *Musical Performance*, Computational Music Science,
DOI 10.1007/978-3-642-11838-8_10, © Springer-Verlag Berlin Heidelberg 2011

integral curves of $\mathfrak{Ts}$. But what is an integral curve of a vector field? Suppose we take a point X of the underlying parameter space $\mathbb{R}^{P.}$. Then there is a unique curve $\int_X \mathfrak{Ts} : J \to \mathbb{R}^{P.}$, defined on an open interval J of the real number line $\mathbb{R}$, such that $\int_X \mathfrak{Ts}(0) = X$ and $d\int_X \mathfrak{Ts}/dt(t) = \mathfrak{Ts}(\int_X \mathfrak{Ts}(t))$ for all $t \in J$, and such that the curve cannot be extended to a strictly larger domain of parameters t. So the curve's tangent at the curve point for curve parameter t is the given vector field at the curve point for that parameter. Intuitively speaking, if one imagines the vector field as being the velocity field of a water stream, the integral curve is the curve a small boat follows when floating on the water.

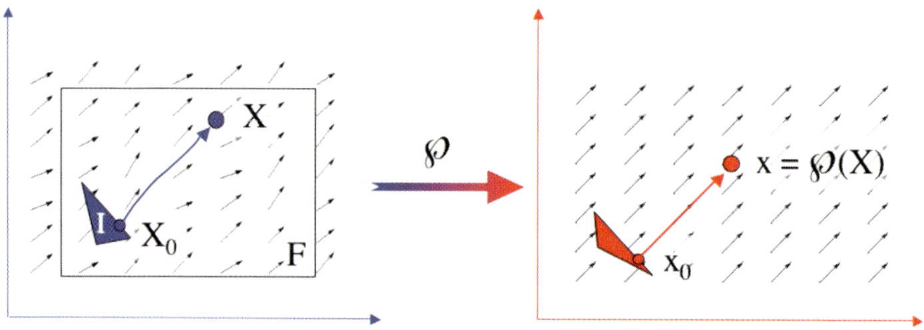

Fig. 10.1. Performance is defined by a field $\mathfrak{Ts}$, defined on a frame F, and an initial performance map defined on an initial set I.

Since an integral curve $\int_X \mathfrak{Ts}$ is the inverse image of the corresponding integral curve $\int_x \Delta$, $x = \wp_{P.}(X)$, the time that elapses when moving on integral curve $\int_X \mathfrak{Ts}$ between two points is the same as the time that elapses on $\int_x \Delta$ between the corresponding points. Therefore, if X_0 is a point on $\int_X \mathfrak{Ts}$ at time t, and if we know the "initial performance value" $x_0 = \wp^I_{P.}(X_0)$ on $\int_x \Delta$, then the performance of X is given by

$$x = \wp_{P.}(X) = x_0 - t\Delta.$$

We may therefore calculate the performance of a point X by taking the integral curve though that point, seeking a point X_0 on the curve whose initial performance is known, calculating the time that elapes from X to that point X_0 and then applying the above formula. We therefore need this data to calculate performance:

1. a performance field[1] $\mathfrak{Ts}$,
2. defined on an n-dimensional cube $F = [a_1, b_1] \times [a_2, b_2] \times \ldots [a_n, b_n]$, Cartesian product of n closed intervals, $[a_i, b_i], a_i \leq b_i$, called the *frame of the performance*,

[1] $\mathfrak{Ts}$ must be a Lipschitz field, see [84, Chapter 33.2.2] for such technicalities.

3. an *initial set* $I \subset F$, where
4. an *initial performance map* $\wp^I_{P.} : I \to \mathbb{R}^{p\cdot}$ is given.

Our performance is then defined for all points $X \in F$ that can be connected to points of I by integral curves (see also figure 10.1). The set of "notes" $K \subset F$ to be performed should consist of such connectable points, of course. It is called the *symbolic kernel of the performance*. The total information $\mathcal{C} = (\mathbf{Ts}, F, I, \wp^I_{P.}, K)$ is called a *performance cell*.

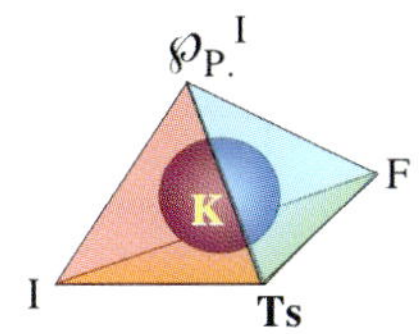

Fig. 10.2. Symbol for a performance cell.

10.2 Initial Sets

It is important to understand the deep meaning of initial performance. We have known performance as a transitional process from mental to physical reality. This is a transition from the score to the acoustical realization, to be archived on sound media such as a CD. Following the valid doctrine—as preconized by Paul Valéry and Theodor W. Adorno—performance is an integral part of the work of art, and this means that, in the sense of communication theory of art as described by Jean Molino, performance is part of the semiosis of the work, and its meaning is not complete except when it is performed.

Fig. 10.3. Sergiu Celibidache (1912-1996).

Put it the other way round: The mental score conveys a part but not the whole content, and only via performance can we complete the work's semiosis. Performance involves a kind of usage of the mental score sign by a performer. More specifically, those signs whose content are not only instantiated but also substantially depend on the user are the well-known shifters. Shifters (also called deictic morphemes) are signs that gain their full meaning only when used. In language, the most prominent shifters are "I," "here," and "now." The lexical meaning of these signs is incomplete and changes significantly when such a sign is used.

Every human individual using "I" creates the subjective individual portion of the pronoun's meaning. Same for the other two signs: Each usage changes their meaning—each time when I say "now," it means a different moment of presence. This contrasts with lexically determined signs, such as "cat," whose complete meaning can be looked up in a dictionary. Performance of a mental score is such a shift from lexicality to full-fledged meaning, since the pure score is essentially less than the work of art. In other words, performance is what semiotics calls a shifter characteristic of the score semantics.

Production of full-fledged meaning is only possible by means of performance, and this adds a semantic aspect to the sign that is a non-trivial function of the performer(s).

The shifter nature of performance is especially acute in initial performance, since this is where the fictitious reality of the score is anchored in reality, namely by the initial value of initial symbols. All the rest is defined by integration of performance fields, but these are only meaningful to physics via initial anchorage. This existential aspect was promimently stressed by the celebrated Romanian conductor Sergiu Celibidache.

10.3 Measuring Performance Fields

Although the formalism of performance cells is a perfect conceptual tool, whether and how it can be applied in practice is far from evident. There is a number of problems that are related to such far-out mysteries. First, the performance field of a normal parameter space with onset, pitch, loudness, and duration is four-dimensional and hard to visualize as such, so lower-dimensional images will be required if possible. Second, even if a representation has low dimension, it is difficult to visualize vector fields with strong intuitive expressivity. Third, if we are given a performance, how can we calculate and the visualize its performance field(s)?

It is clear that such questions are extremely important for performers, for their instructors, and for the empirical research on existing performances. The solution to this problem has been offered in collaboration with my computer science students Stefan Göller and Stefan Müller [86]. They have programmed a software component, called *Espresso Rubette*, of the music software environment RUBATO®. The component has a 3D interface that looks like an espresso machine (figure 10.4). It takes a MIDI file of a composition and the MIDI file of its performance, compares them and generates a corresponding performance field. The comparison is a highly non-trivial task since one has to match the score events with the performed events. These may be wrong by the musician's errors, or there might even be ghost events played by error but not corresponding to any written notes. The performed events are also, by the very action of performance, not in the same relative position as written on the score: The onset times of notes written in a chord may differ by slight arpeggi, and the duration will be articulated.

There are a number of rather good matching algorithms, but none of them is absolutely reliable, also because there is no a priori reason to have a perfect match in view of the mentioned errors and deformations. The program's second task, after a match has been found, is the construction of a performance filed. Here there are two subtasks: finding a finite number of "representative vectors" of the field from the finite number of events available from the given

Fig. 10.4. The *Espresso Rubette* component of RUBATO® takes a MIDI file of a composition and the MIDI file of its performance, compares them and generates a corresponding performance field, which is visualized as a color field by use of the color circle.

composition and interpolating these vectors to a field that is defined everywhere on the composition's frame. The interpolation task is trivial. The vector field can also be represented as a color field matching the vectors' directions with positions on the color circle; the vctors' lengths determine the color intensity. Figure 10.4 shows such a visualzation of a part of the Czerny exercise shown as input to the *Espresso Rubette*. The black points on the color field to the right are the performed note events. The construction of the representative vectors is quite subtle, but it can be done on the basis of standard methods of linear algebra. The point here is to find good pairs of vectors that describe the Jacobian matrixes, where "good" means that the selection must cope with robust positions of the transformations' base vectors, see [86] for details. It is evident that such a software is the germ of a revolutionary tool for performance education because the student can play a piece on a MIDI piano, and—while the performance is ongoing—the color field on a big screen shows immediately the performance field as a common reference for the instructor and the student, a tool which enables a detailed, undelayd, and objective reference for the delicate pedagogical work of teaching refined performer artistry.

11

The Category of Performance Cells and Hierarchies

On trouve toujours l'homogène à un certain degré de division.
Paul Valéry [142, I, p.209]

This chapter completes our study of the structure of performance. Recall from chapter 5 that we had four local-global dichotomies for performance theory: instrumental, parts, dimensions, and evolution. We are not going to discuss the first two, but dimensions and evolution will be dealt with. Dimensions are what we want to discuss now, and evolution will be discussed in chapter 21.

We have defined the minimal units that are full-fledged data for the construction of performance, namely performance cells $\mathcal{C} = (\mathbf{Ts}, F, I, \wp^I_{P.}, K)$. But we have also seen that certain parameter spaces do not need other parameters in order to define performance of *these* parameters. The two examples we have dealt with are the default performance of onset and pitch on one side:

$$\mathbb{R}^E \xleftarrow{pr_E} \mathbb{R}^{EH} \xrightarrow{pr_H} \mathbb{R}^H$$

and the performance of onset and duration on the other:

$$\mathbb{R}^{ED} \xrightarrow{pr_E} \mathbb{R}^E$$

G. Mazzola, *Musical Performance*, Computational Music Science,
DOI 10.1007/978-3-642-11838-8_11, © Springer-Verlag Berlin Heidelberg 2011

These arrows can now be made more precise when relating them to performance cells. This works as follows: Suppose in the first situation that we are given a performance cell $\mathcal{C} = (\mathbf{Ts}, F, I, \wp^I_{EH}, K)$ for onset and pitch. And also suppose that the field $\mathbf{Ts}$ is a Cartesian product $\mathbf{Ts} = \mathbf{Ts}_E \times \mathbf{Ts}_H$ of an onset and a pitch field, i.e., tempo and intonation. Then we can forget about pitch when performing onset: Tempo is independent of intonation, and vice versa. Take now the projection $K_E = pr_E(K)$, $K_H = pr_H(K)$ of the symbolic kernel to onset and pitch, respectively. Also suppose we have "reasonable" initial sets I_E, I_H and initial performances $\wp^I_E, \wp^I_H$ in onset and pitch (we are imprecise here, but this is a technical subtlety that would disturb our understanding), and take the projections $F_E = pr_E(F), F_H = pr_H(F)$ as onset or pitch frames, respectively. We therefore get two performance cells: $\mathcal{C}_E = (\mathbf{Ts}_E, F_E, I, \wp^I_{EH}, K)$ and $\mathcal{C}_H = (\mathbf{Ts}_H, F_H, I, \wp^I_{EH}, K)$. Then the projections p_E, p_H can be viewed as "morphisms" (a kind of generalized map) between these cells, yielding this diagram:

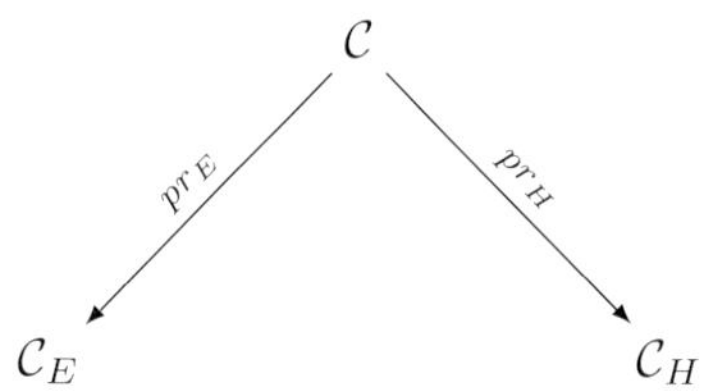

Similarly, we can generate an arrow of performance cells in the second case:

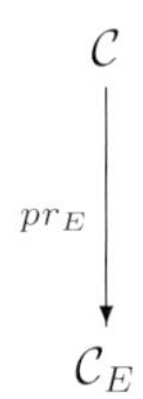

in which case there is no arrow to a cell in the space of durations, since with no performance fields here, no such cell is possible.

The general definition of such a morphism between performance cells is this: Take two cells $\mathcal{C}_1 = (\mathbf{Ts}_1, F_1, I_1, \wp^I_{P.}, K_1)$, $\mathcal{C}_2 = (\mathbf{Ts}_2, F_2, I_2, \wp^I_{Q.}, K_2)$, where $Q.$ is a subset of parameters of the set $P.$ and consider the projection $p : \mathbb{R}^{P.} \to \mathbb{R}^{Q.}$. Suppose these conditions are satisfied:

1. $p(F_1) \subset F_2$,
2. $p(K_1) \subset K_2$,
3. $p(\mathbf{Ts}_1) = \mathbf{Ts}_2$, which means that the components of $\mathbf{Ts}_1$ in $\mathbb{R}^{Q.}$ do not depend on parameters other than those in $\mathbb{R}^{Q.}$ and have the values of $\mathbf{Ts}_2$;

plus some technical conditions on the initial performance that we omit here (but see [84, Chapter 35.2]). These data define what we call a *morphism of performance cells*. It is denoted by

$$p : \mathcal{C}_1 \to \mathcal{C}_2.$$

To be clear, the underlying map of a performance cell morphism is always a projection of parameter spaces. The only serious point that turns it into a morphism is the set of the above three conditions (plus those technical conditions). With this conceptual architecture, we can now define a *performance hierarchy* $\mathcal{D}$ as being a diagram $\mathcal{D}$, whose vertexes are performance cells and whose arrows are morphisms of performance cells. So the above diagrams give rise to simple performance hierarchies.

The advantage of this hierarchical representation of performance is two-fold. First, it eases calculations for parameter spaces that are projections of higher-dimensional ones. It is much easier to calculate, for example, three one-dimensional values than one three-dimensional one since the calculation of higher-dimensional cases requires the numerical integration of vector fields, which is equivalent to the numerical solution of ordinary differential equations (ODEs). Second, for the genealogical theory of performance, i.e. the theory that describes performance as an unfolding process, starting with the unshaped prima-vista rendition and ending with the artistically detailed shaping, one needs such hierarchical structures in order to derive more sophisticated performances from simpler ones.

In order to show concrete situations, we give in figure 11.1 an example of a performance hierarchy for the piano. It is the default hierarchy, i.e. a hierarchy defined for the most simple configuration of the piano. It is the starting points for more sophisticated performances.

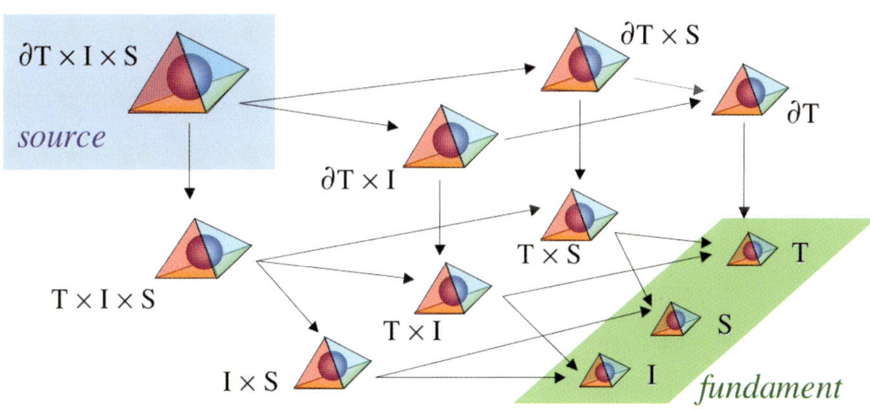

Fig. 11.1. The piano hierarchy spans between the source cell with all four piano parameters—onset, pitch, loudness, and duration—and the three fundamental cells for onset, pitch, and loudness, respectively.

It lives in the four-dimensional space $\mathbb{R}^{EHLD}$ and is realized in the source cell in figure 11.1. This cell has the Cartesian product of three fields: the one-dimensional fields I for intensity (dynamics) and S for intonation, and the two-dimensional parallel field ∂T of articulation. This cell projects onto three three-dimensional fields $T \times I \times S$, $\partial T \times I$, and $\partial T \times S$, where the first projections stem from the projection of the parallel field onto tempo. The triple Cartesian product cell for $T \times I \times S$ projects onto three two-dimensional cells $I \times S$, $T \times I$, $T \times S$, whereas both, $\partial T \times I$ and $\partial T \times S$ project onto the parallel field ∂T. All these two-dimensional cells project onto the one-dimensional cells I, S, and T of the fundament. When refining this default performance scheme, one will have to act on some of these cells and thereby define a new hierarchy that as a matter of fact will have fewer vertexes because the independence of parameters will be deranged. A more complex default hierarchy for the violin has been described in [84, Chapter 35.3.3].

Part III

Expressive Theory

12

What Is Expressive Theory?

This last album is not titled as a memorial album or as an album in tribute because it was titled by Coltrane himself the Friday before his death on Monday, July 17, 1967. He and Bob Thiele were considering words that might apply to the sense of this album, and finally Coltrane said, "Expression. That's what it is."

Nat Hentoff [51]

Expressive theory has been sketched in sections 4.6, 4.7, 4.8 of chapter 4 dealing with (oni)ontology of performance. We have seen that expression splits into semantics and rhetorics: What is expressed, and how this is performed. Performance theory as it stands now partitions this complex of expressivity into three big themes: emotion, gesture, and analysis. These refer not to the rhetoric aspect, but to the variety of contents that are transferred to the audience, i.e. to the two axes of realities and embodiments on which these contents are distributed. We have given a sample of such contents in section 4.7, especially in figure 4.8.

Although these three themes are not strictly identifiable with coordinates on the axis of realities, one is used to relate them in a first approximation, namely emotion to psychology, gesture to physics, and analysis to mental/symbolic reality (figure 12.1). These relations are plausible however because emotion is mainly psychological. But its gestural aspect in the sense of the etymological interpretation of emotion as "ex-motion", moving from inside out, would refer to gestures. Also it is true that to the naive understanding, gestures appear as physical utterances, but in a more sophisticated approach they may prominently also refer to gestural movements in symbolic spaces, a situation encountered in the gestural analysis of motivic shapes, for example. And third, analysis is evidently related to symbolic structures, such as harmony or counterpoint, but it can deal with physical structures, such as sound anatomy or gesture syntax.

G. Mazzola, *Musical Performance*, Computational Music Science,
DOI 10.1007/978-3-642-11838-8_12, © Springer-Verlag Berlin Heidelberg 2011

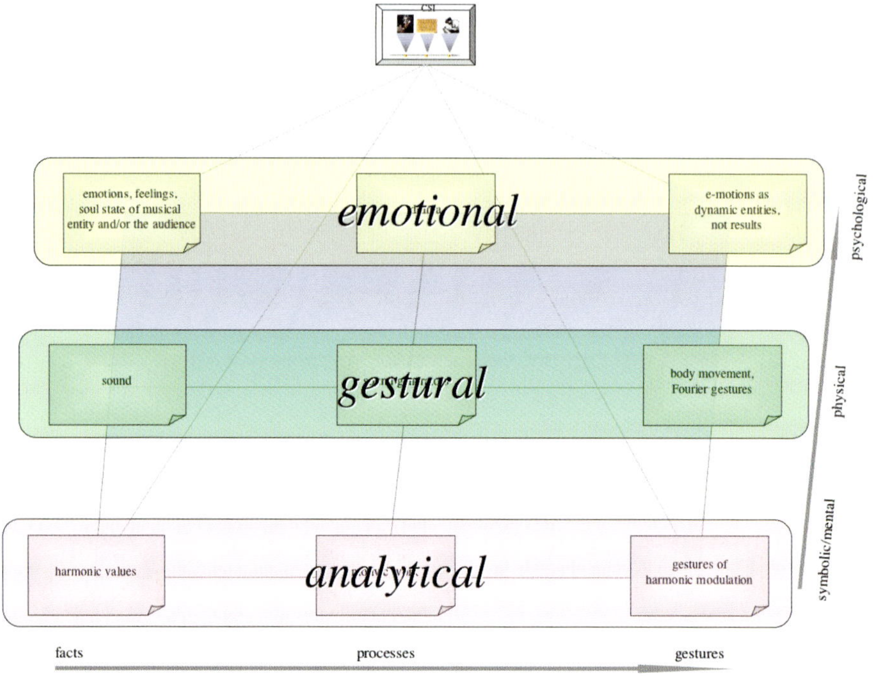

Fig. 12.1. The partition of semantic positions into expressive fields associated with emotion, gesture, and analysis.

Once we agree on this partition of semantic expressivity, it turns out that the correlated rhetorical expressivity may vary in a dramatic sense. That is to say that each of these three expressive CSI positions asks for categorically different rhetorical strategies to reach the ISC position in an effective way. Performing emotional contents requires rhetorical tools that would enable the audience to perceive and identify emotional displays, and then to deal with such contents in a culturally encoded manner. And gestural expressivity requires a rhetoric of gestures that fits in the audience's imaginations. This is extremely difficult since the audience are usually not moving their bodies, not dancing (unfortunately, since that would ease much of the understanding). Even dancing with Schönberg's String Trio op.45 can reveal and generate an infinity of understanding not at reach when just sitting there and listening, but see [91, Chapter 9.3]. Analytical expressivity looks easy since the analytical structures are clearly identifiable. But the problem here is that there is no canon of rhetorical devices for such expressivity. How should one express a cadence? What about the expression of a tension between left and right hand rhythmics, as it is the case in Schumann's famous *Träumerei* op.15/7. We shall discuss this example in detail in section 16.1.

12.1 Experiments in Expressive Performance

The conceptual setup of expressive theory might be acceptable from the epistemological point of view, but there is a huge problem relating to the adequacy of rhetorical expression for given semantic expressivity. To put it in simple terms: How can we know that—and if yes, to what extent—a rhetorical device is effective in transmitting the message? Can such a question be answered by a priori categories of aesthetics? Or do we have to make experiments, by trial and error?

To be clear, we are not taking any of those ideological positions about beauty and style and what not. We are not dealing with New Music propagation or invocation of would-be eternal values of harmonies. Our concern is the relationship between the intended contents, their rhetorical transfer, and then the audience's perception and reception thereof. Of course, it might and it will happen that the ISC positions in question are heavily affected by such ideological loadings. But this is their problem and must be taken into account when evaluating the situation. It is in particular not adequate in a valid performance theory to claim the *deus ex machina* of music psychology, that fictitious being the *expert listener*, which has been invoked by Ray Jackendoff and Fred Lerdahl in their celebrated *Generative Theory of Tonal Music* [53]. This invention is nothing less than a hidden instantiation of prejudices about hearing in the form of a fictitiuous expert who pretends to know what to hear and how to listen.

We shall see in the following exposition of expressive approaches to performance that such ideological positions are taken—sometimes unconsciously, sometimes explicitly—in order to spread a specific *Weltanschauung*. But there is also a hard problem behind such decisions: What would one want to do to test adequacy of expression beyond simply propagating a specific solution without knowing what is the outcome? And it happens that a theoretical hypothesis turns out to be false when it comes to test it in realistic contexts. A simple example is a test concerning appreciation of common versus difficult art music, which Scott Lipscomb and I have conducted at our School of Music [70]. We played three versions of such music: without visuals, with abstract visuals, and with the video showing the musicians in their physical performance. We believed that the video would enhance the level of appreciation of difficult art music. Our belief was based on the hypothesis that the embodiment of music and its gestural dynamics would help listeners understand the complex musical shapes. This was plausible from a general gesture-theoretical approach to performance also supported by Theodor W. Adorno's theories (see section 14.3).

But the result was the opposite. It seems as if the physical presentation of such performance would even make appreciation more difficult. The abstract visual strongly enhanced appreciation of difficult art music, whereas viewing musicians playing generated the mimimal appreciation. We could however conjecture why this paradoxical result was possible. Our test population was taken from students of a pop music class, which means that these individuals were probably not familiar with art music. Therefore, being exposed to this kind of art was not an a priori agreeable experience to them. This fact was not neutralized by showing the musicians at work since it intensified the already negative auditive experience. It seems that the presentation of the performing musicians resulted in a close-up on an already disagreeable experience. Opposed to this situation, the abstract visuals gave the students the chance to unfold their own imaginations and to disconnect from the disagreeable musical input. Future experiments will hopefully reveal more details about this astonishing outcome.

Fig. 12.2. Reading Aristotle's physics versus doing Galilean experiments (extract from a painting by Giuseppe Bezzuoli).

We can conclude that one must make experiments with expressive performance in order to learn what does and does not work. This consequence is not so easy to realize! How would one shape performance in different ways according to given analytical semantics? Which is the variety of tempi to be shaped around a given cadential harmony? Besides the mere question of what would be the shaping, it is also the question of which performer or which tool would be capable of shaping such a variety. Human performers might produce excellent or at least unique performances. But the performance with exactly this specific tempo curve? Who can do this, and repeat it identically, or just everything 93 percent of the previous tempo curve? The fact is that one needs to think about how to make experiments with expressive performance. Experiments must be repeatable identically, a very difficult if not impossible condition upon human performers.

The consequence of these thoughts is that either there is no experimental performance science or we must invent tools that allow for precise design and identical replication of expressive performances. Looking back in history, it becomes clear why performance research has taken so long to become a full-fledged science: There were no such tools as needed for valid experiments. Performance theory lacked sufficient experimental counterparts. Theoretical physics suffered from the same deficiency before Galileo started doing replicable precise experiments instead of reading books about Aristotelian speculations. So this is the background for a historical localization of what is happening now: the foundation of an experimentally anchored expressive performance theory, a science that uses computers and sound synthesizers or cutting-edge player piano technology to construe performances according to precise rhetorical rules. We shall come back to this issue on several occasions in this book, most prominently in Part IV, dealing with the RUBATO® software for analysis and performance in music.

13

Emotional Expression

The emotional virus lives and thrives in the gap between expectations and perceived reality.
Doc Childre and Bruce Cryer

Before we discuss prominent approaches to emotional expressivity in performance, we should introduce the very concept of emotion as it is understood in psychology, together with some remarks on neurological correlates to emotion.

13.1 What is Emotion?

In their book *Music and Emotion* [56, p. 71], John A. Sloboda and Patrick N. Juslin give a concise description of the study of emotions in music:

> *Psychology is concerned with the explanation of human behaviour. Behaviour includes overt action as well as 'inner' behaviour, such as thought, emotion, and other reportable mental states. It can include behaviour of which the agent is not fully or even partly aware, such as the dilation of the pupils of the eye. (...) A psychological approach to music and emotion therefore seeks an explanation of how and why we experience emotional reactions to music, and how and why we experience music as expressive of emotion.*

Most importantly, they distinguish those emotions that are reified in the audience from those emotions that are presented from the artist's and or composer's side. In our scheme of expressivity, we did not distinguish between these two types, but it was meant that the expressed emotional content could be either perceived or experienced by the audience.

G. Mazzola, *Musical Performance*, Computational Music Science,
DOI 10.1007/978-3-642-11838-8_13, © Springer-Verlag Berlin Heidelberg 2011

The critical concept here is "emotion": desire, love, jealousy, sadness, calm, satisfaction, admiration, or curiosity, for example. What is emotion in psychology? Emotions are characterized in [56, p. 74] by three areas:

Characteristics	Examples
self-report	feelings, verbal descriptions, checklists, rating scales, etc.
expressive behavior	facial expressions, gestures, vocalization, etc.
physiological measures	blood pressure, skin conductance, muscle tension, EKG, EEG, etc.

This tabular scheme is not a definition in the strict sense but a list of phenomenological properties associated with emotion. In the first row, self-report describes the access to emotion via reports given by the human that experiences emotion. The distinction between emotion and feeling is that emotion has an inner processuality, it is not just an amorphous thing. I would call disgust or desire a feeling, because they have no inner logic. In contrast, sadness involves a rather complex configuration of objects or persons we have lost and to which we were related in a strong way because of a number of other factors, and now all those components are put into question, etc.—it is an entire narrative that causes the emotion of sadness. So emotions can be described in a narrative of cognitive character. Emotions may refer to feelings but are not reduced to such. Feelings pertain to what is known as "qualia," a philosophical term invented by Charles Sanders Peirce and then spread by Clarence Irving Lewis [69], meaning those phenomena of human existence that are intrinsic, private, ineffable, and directly perceived. Qualia are, for example, seeing a color, smelling a flower, being nauseated. In clinical medicine, the qualia of pain are reduced to numbers, from zero to ten, but evidently not described as what they are.

In the second row, expressive behavior circumscribes what we do when experiencing emotions, how we move, and what gestures, postures, facial expressions or vocal utterences we shape. It is the audio-visual level of embodying emotions. This is perhaps the most faithful representation of the etymological root "ex movere," to move out, of the word "emotion."

The third row, physiological measures, refers to the changes in our body state induced by emotions, such as rising blood pressure or increasing heart rate. Remarkably, all except the first row (self-report) could also happen to animals. The lack of language might be the critical difference in that only humans are capable of developing that narrative of cognitive processes required for a self-report and of the self-consciousness required for the delivery of a self-report.

Emotions are classified by essentially two methods: categories, following the work of Richard S. Lazarus et al. [67], and the Circumplex Model, following the work of James A. Russell et al. [117]. The first method establishes a small discrete list of basic emotions:

Emotion	Juncture of plan	Core relational theme
Happiness	Subgoals being achieved	Making reasonable progress towards a goal
Anger	Active plan frustrated	A demeaning offense against me and mine
Sadness	Failure of major plan or loss of active goal	Having experienced an irrevocable loss
Fear	Self perseveration goal threatened or goal conflict	Facing an immediate, concrete, or overwhelming physical danger
Disgust	Gustatory goal violated	Taking in or being close to an indigestible object or idea (metaphorically speaking)

The advantage of this approach is that it offers narrative rationales to describe the processual nature of these emotions. There is a plan with goals and subgoals, and there is a logic of how the plan's success or failure will entail one's corresponding experience (loss, danger, progress, etc.). It claims that all other emotions are derived from these, be it by blending some of the basic ones or by some mechanism of specialization.

This approach has been called speedy rather than precise. Clearly its strength, namely the small number of core emotions, is also its weakness, since the theory only works if the mechanism for the construction of derived emotions is made explicit, and if there are rationales why these and not other emotions would be basic. Interestingly, only one of the above basic emotions can be called positive: happiness. The others are utterly negative. The model is visibly deduced from biological criteria of the strugge for survival. But in music, and not only in this field of human culture, this setup looks quite inappropriate. Why not love and hate, attraction and repulsion, fear and trust? And it is also not clear why disgust is called an emotion and not a feeling. To invoke a "gustatory goal" looks like an artificial casting of this feeling into the theoretical scheme.

The other approach to classification of emotions is termed "Circumplex Model" and works with a geometric parametrization of emotions (figure 13.1). Every emotion is represented by a point in the two-dimensional space spanned by the axis of pleasantness (horizontal) and the axis of activation (vertical). So the underlying hypothesis is that all emotions are a mixture of pleasantness and activation. For example, inspiration, desire, and love have two positive coordinates and therefore lie in the first quadrant of the display. The advantage of this system over the above system of categories is that by its geometrical approach, it enables an infinity of emotions. The strategy is simple: Emotions are arithmetical mixtures of two basic components, which need not be emotions, because they are just the underlying ingredients. The geometric approach also has the automatic structure of polarity, namely the reflection of an emotional point (x, y) at the origin, yelding its "polar" antipode $(-x, -y)$. Whether every emotion has such an antipode that is also attributed to an emotion is an open question, but a number of polar pairs, such as "disappointment" versus "fascination" or "happiness" versus "sadness," do present examples of such geometric transformation.

The geometric flavor of this model automatically raises questions concerning the implications of the geometric methods on the understanding of emotions. To begin with, the distance between emotions *qua* points will mean something. How can it be determined whether this matters? Or is it just an arrangement of ordered nature, with the emotional "points" qualified only by the total order of their relative positions to one another? Second, it is not evident why we should have only two basic dimensions. And why activation and pleasantness? There have been other proposals in this spirit, sometimes also three-dimensional, but the question subsists if we are not given an explicit construction method. The mechanism underlying this geometric parametrization however remains obscure.

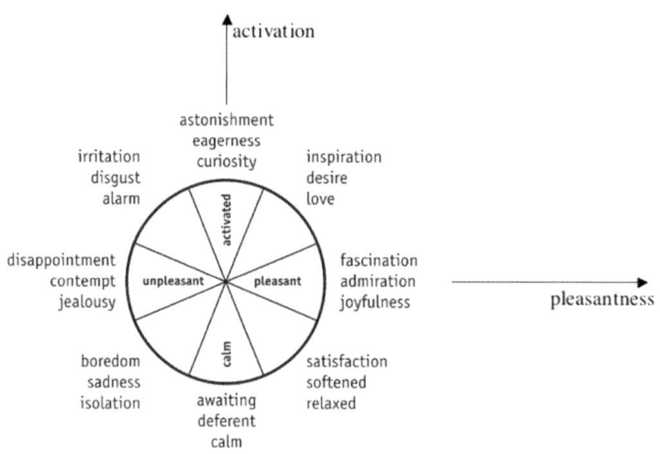

Fig. 13.1. The two-dimensional Circumplex Model of James A. Russell et al. inscribes emotions in a two-dimensional space spanned by the axis of pleasantness (horizontal) and activation (vertical).

13.2 Some Physiological Evidences

We should now give a number of empirical evidences of the neurophysiological emotion-related responses to music, so we are relying on the third characteristic, physiological measures, in the above definition of emotions. A number of neurophysiological experiments have been done where the electrical activity of the human brain was investigated in its response to acoustical inputs of musical structures. The first two of them have revealed significant differences between female and male listeners. In 2003, Stefan Koelsch et al. [61] measured event-related brain potentials (ERP, short event-induced spikes delayed by 300 to 600 msec) taken from electrodes on the surface of the skull (figure 13.2). The chosen population were 5- to 9-year-old girls and boys without musical training. The musical stimuli were three short cadential sequences of chords, each in two variants. The first was the typical $I - IV - V - I$ cadence, with variant $I - IV - I - V - I$. The second was altered to take the neapolitan chord instead of the last I, or the middle one in the variant, and the third showed a cluster instead of the final I or the middle I in the variant. The distribution of the ERP over the skull is shown to the right in figure 13.2. We see that the girls have a symmetrical activity map, whereas the boys show a maximum of activity in the right frontal brain[1].

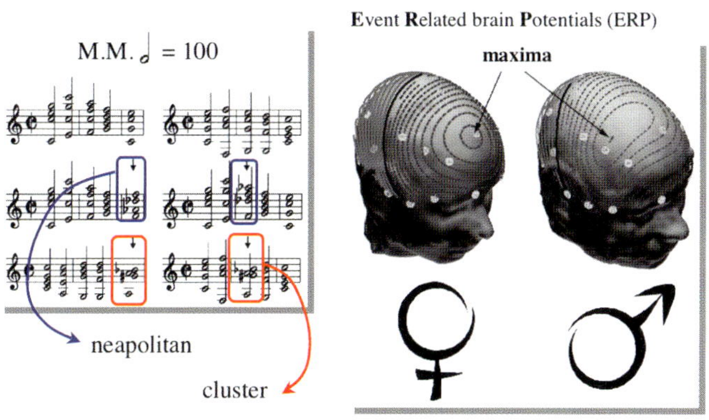

Fig. 13.2. Stefan Koelsch's experiment with event-related potentials as a response to three cadential chord sequences.

In his extensive studies of EEG responses to musical stimuli [107] from 1986 to 1998, Hellmuth Petsche et al. obtained quite detailed information about the localization and coherence of surface EEG activity in the different classical frequency bands, namely $\theta = 4 - 7.5$ Hz, $\alpha = 8 - 12.5$ Hz, $\beta_1 =$

[1] We are aware that the big and important topic of gender in musical composition, performance, and theory is not dealt with in this book. This has two reasons: To begin with, the field would deserve a very cautious discourse on the different rationales for gender differences in music. Since this book is an introduction to performance theory, we do not delve into this complex field. We are aware however that at present, the sensibility for embodied aspects of performance is more evolved with female performers. We refer to the contribution by saxophonist Lisa R. Rhoades in section 18.3.1 as a proof of this sensibility. Second, it is a deplorable prejudice that female composers write girlish music.

13 – 18 Hz, $\beta_2 = 18.5 - 24$ Hz, $\beta_3 = 24.5 - 31.5$ Hz. His brain maps show power ("Leistung"), local coherence ("lokale Kohärenz"), and interhemispheric coherence ("interhem. Kohärenz"). Coherence means that they measure how strongly electrical activities in different localities of the brain are coupled to each other, and this is a sign for simultaneous, and therefore coordinated, processing of musical stimuli in different brain areas.

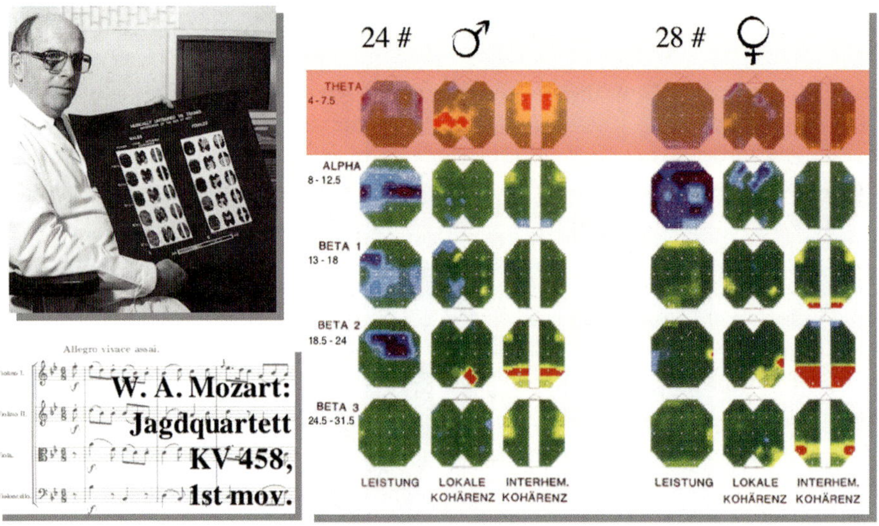

Fig. 13.3. Hellmuth Petsche's experiments with EEG responses in different frequency bands, for male and female populations, to the first movement of Mozart's string quartet KV 458.

Petsche presented the first movement of Mozart's string quartet KV 458 to 24 male music students and to 28 female music students. So the response has to be interpreted as a global reflection of the musical structure. The results (figure 13.3) show that interhemispheric coherence, in particular in the beta band, which is correlated to intellectual brain tasks, is much stronger (red color) for females than for males. This confirms the well-known fact that females have a stronger general interhemispheric activity than males. We also see the significant local coherence in the right hemisphere on the β_2 band, which confirms that the right hemisphere is an area of holistic music processing, typically needed for the recognition and evaluation of complex shapes, like motives and melodies. The θ band is neglected because it can be affected by signal noise.

The findings of Petsche are quite sensational in view of the question about the nature of cognitive music processing in the right cortical hemisphere. Since Petsche's findings confirm cognition and evaluation of complex shapes via local coherence in the right hemisphere, we might ask what happens if this coherence is impaired by some physiological conditions. One of those impariments is reported from anatomical measurements of brain weights in schizophrenics. It has been shown that these patients suffer from a significant loss of axon mass in the right hemisphere [76, Chapter 1], axons being the connectors between neurons. This implies that the local coherence in the right hemisphere is lower than for normal humans. This has been confirmed by EEG investigations similar to Petsche's coherence measurements. The labyrinthic existentiality of schizophrenics appears as a consequence of the thinned-out network of neurons: They have to run through a labyrinth of axons instead of performing on a dense neuronal maze. The linearization or reality is evident in the art of schizophrenics, such as the graphical music work of Swiss schizophrenic Adolf Wölfli (figure 13.4).

Fig. 13.4. Swiss schizophrenic Adolf Wölfli has produced a large number of drawings of musical topics, mainly score-like displays. The photography shows him with one of his paper trumpets.

All of this should also have consequences for the relation of schizophrenics to music. It would imply that they have a lowered performance in creating, recognizing, and processing complex musical shapes. In particular, this would suggest that there are no great schizophrenic composers, who are required to imagine complex shapes while creating high-ranked works. It is in fact true that no such composers are known. The only critical case was Robert Schumann, but it has been proved that he was not schizophrenic but suffered from a tumor on the skull base [126]. Other psychoses, such as depression, do not affect musical creativity, as is, for instance, beautifully and tragically shown by the example of Peter Tschaikowsky. Of course, it is by no means a negative judgment about schizophrenics to put into question their compositional abilities. It might also be true that in future times they could compose a type of music that has high qualities that we cannot view and appreciate at present.

A third example of neurological evidence of musical stimuli, and this time strongly related to emotion, is my joint research with epileptologist Heinz-Gregor Wieser at the University Hospital in Zürich from 1984 to 1986 [74], [149], [150]. We were in the interesting position to have depth electrodes implanted in humans with chemically intractable focal epilepsy for presurgical evaluation (localization of the focus). During this evaluation, it was possible to present musical structures to the patients via headphones and to have the depth structures of the brain respond to this input. For medical reasons, a number of electrodes were positioned in the hippocampal formation of the limbic system,

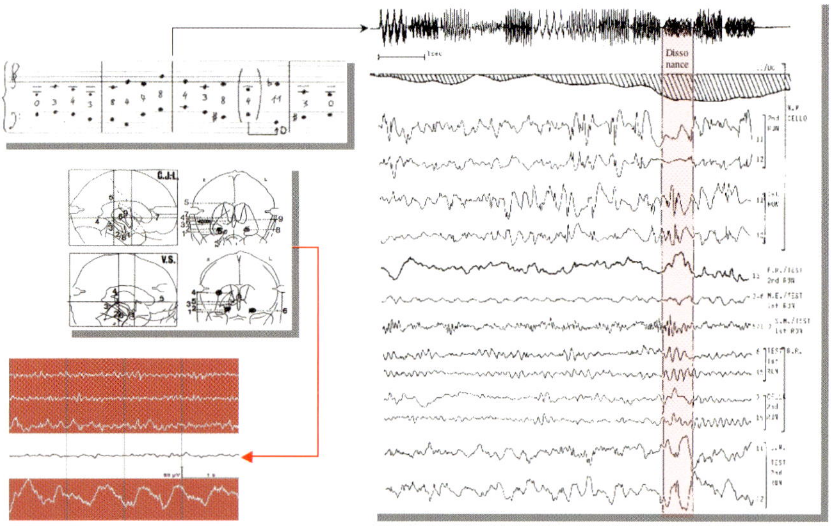

Fig. 13.5. The depth EEG recordings of a counterpoint example with defective interval (dissonant major seventh replacing the major third consonance). The depth EEG from the hippocampal formation of the emotional brain shows a significant disruption when the dissonance appears.

which is a prominent structure of the emotional brain (situated below the neocortical brain layers). Figure 13.5 shows the electrodes in X-ray imaging and the EEG derived from these electrodes in the range of some $50\mu V$. We see the input score, which is an example of first species Fuxian counterpoint, on which we have changed one of the consonances (a major third) to a major seventh, a strong dissonance in the third but last position. The sequence of a number of EEG waves reacting to this dissonant disruption in the consonant context is shown to the right. We see a significant change in the EEG structure when the dissonance appears. This proves that beyond the conscious perception of dissonances versus consonances, this EEG disruption is produced in the deep structures of the emotional brain. The patients in these tests were normal European male adults with common musical taste and no instrumental education. *The result of this investigation is that in the deepest structures of the emotional brain (in the archicortex positioned below the neocortex), basic musical structures of consonances and dissonances are significantly distinguished from each other on the level of EEG waves.* We therefore have strong arguments for the emotional effect of music beyond the conscious self-report or behavioral response.

13.3 Manfred Clynes' Essentic Forms

Manfred Clynes is an Australian pianist and music theorist who in the early 1980s pronounced very critical opinions about the nature of musical scores [16]:

> *In Western culture we have devised a singular means of killing music—writing it down in a score. It then has to be resuscitated or resurrected in performance. The performer has to supply all the nuances, the microstructure that was not and could not be notated by the composer, in order to bring the music to life. Therein lies his art.*

Fig. 13.6. Manfred Clynes.

This radical insight is not an isolated one—we shall see that Theodor W. Adorno shares it without reserve. The interesting point with Clynes, whom I have known as a very gentle person, is that he not only complains, but also offers a theory of how to overcome this amputation of music by scores. His approach is based on the hypothesis of emotional semantics in music. The signs carrying such semantics are called *essentic forms.* Clynes argues that they are biologically programmed into our nervous system and therefore music obtains its universal function. The emotional meaning of music refers to an activation of essentic forms. The musical shape of such forms is expressed in mainly agogical and dynamical performance structures. Clynes exhibits a number of such forms, also referred to as "microstructure" or "pulse," which are organized hierarchically and go from intra-note shapes, dealing with the single sound's envelope, through short four-note forms and up to forms comprising measures, phrases, and higher syntactical units.

The typical essentic form described by Clynes is the four-note pulse. He claims that every composer has his/her own pulse; see the original representation of such pulses for Beethoven, Mozart, Haydn, Schubert, Schumann, and Mendelssohn in figure 13.7. Clynes refers to the performance of a four-note motif. The nominal duration and loudness of the notes is 100, and the deviation is shown in the first and second line, respectively. The same information is shown in graphical notation to the right, where the rectangles' widths represent duration and their heights represent loudness. The connecting curve of the upper middle of these four rectangles defines the gestural shape associated with these pulses. We shall come back to this aspect in chapter 14 on gestural expressivity.

This approach seems quite simplistic, since one would expect that for a given composer, many different pulses can be exhibited, and that they also depend on the performer. But the general idea is remarkable because Clynes seems to have been the first theorist to connect the precise performance map to emotional contents.

BEETHOVEN				
Duration	106	89	96	111
Loudness	1.00	.39	.83	.81
MOZART				
Duration	105	95	105	95
Loudness	1.00	.21	.53	.23
HAYDN				
Duration	108	94	97	102
Loudness	1.00	.42	.68	1.02
SCHUBERT				
Duration	97	114	98	90
Loudness	1.00	.65	.40	.75
SCHUMANN				
Duration	96	116	86	102
Loudness	.60	.95	.50	1.17
MENDELSSOHN				
Duration	118	83	95	104
Loudness	1.00	.63	.79	1.12

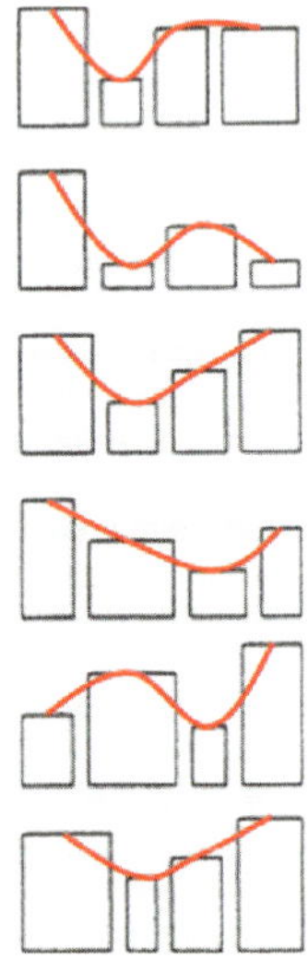

Fig. 13.7. Clynes' pulses for Beethoven, Mozart, Haydn, Schubert, Schumann, and Mendelssohn.

13.4 Reinhard Kopiez' and Jörg Langner's Theory of Oscillating Systems

Fig. 13.8. Reinhard Kopiez (left) and Jörg Langner.

German music psychologist Reinhard Kopiez and German acoustician Jörg Langner have proposed a "Theory of Oscillating Systems" (TOS) that connects performance to emotions on the basis of a system of 120 sinoidal oscillators from 8 Hz down to 0.0008 Hz, which are distributed in logarithmic steps [63], [66]. These oscillators are claimed to exist in the human neuronal cognitive system. Kopiez and Langner apply this oscillator system to dynamical curves of musical pieces in a kind of Fourier analysis that was, however, never published in its precise processing. The hypothesis behind this theory is that the neuronal oscillators are responsible for the emotional validation of music. In the authors' understanding, this neural system is responsible for both perception and production of emotional expression.

As it stands, this approach is very much in the spirit of Clynes' approach, but it makes those essentic forms more concrete under the umbrella of sinoidal decomposition of sounds. No detailed attribution of emotions to the components' loading of the oscillators has been given, and there is no experimental

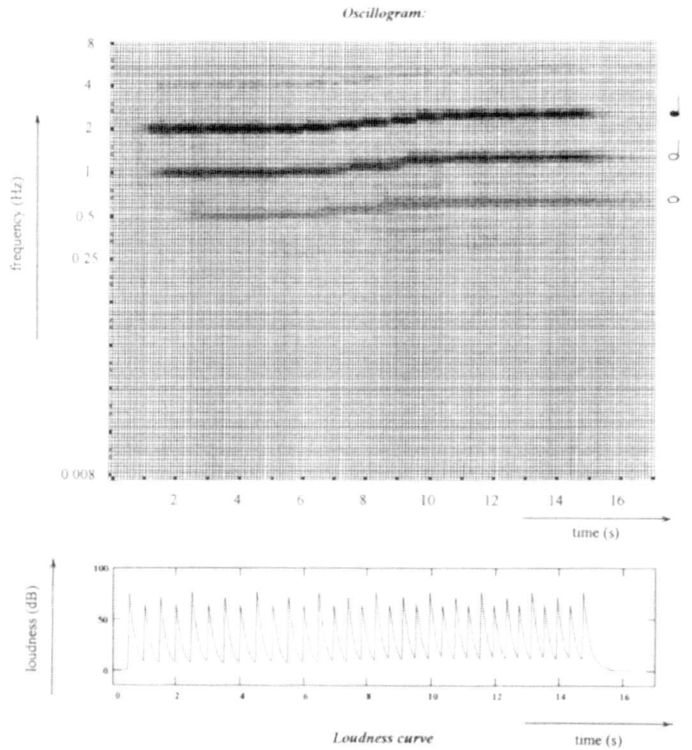

Fig. 13.9. Theory of Oscillating Systems: the oscillogram (above) and loudness curve (below) belonging to quarter notes played in a 4/4-time signature by a drum computer with an accelerando at the end. This accelerando leads to a parallel upward movement of the dark bands in the oscillogram, which means that the activation changes to higher oscillator frequencies.

verification of the existence of such oscillators. Figure 13.9 shows an example of such a TOS structure.

13.5 Anders Friberg's Feature-based Model

So far we have seen theories about emotion being related to musical structures in performance, such as agogics and dynamics with Clynes and the TOS analysis of the dynamics curve with Kopiez and Langner. Both approaches have claimed biological rationales for the evocation of emotions, although this is an overly simplistic understanding of the cognitive narrative mechanisms producing emotions.

The Swedish investigations undertaken by Anders Friberg, his teacher Johan Sundberg (the father of modern performance theory, but see later in chapters 14 and 15), and others [37] is more explicit and concrete than these biologically driven approaches. It uses the performance software *Director Musices* developed by Friberg and Sundberg. This one can shape a number of performance features for computer-generated performance and therefore can be used to shape synthetic performance according to given rules.

Here are the performance features used in this approach:

- Timing: tempo, tempo variation, duration contrast
- Dynamics: overall level, crescendo/decrescendo, accents
- Articulation: overall (staccato/legato), variability
- Timbre: Spectral richness, onset velocity (meaning onset loudness in MIDI terminology)

These are complemented by a number of composer-given features:

- Melody: range (small/large), direction (up/down)
- Harmony (consonant/complex-dissonant)
- Tonality (chromatic-atonal/key-oriented)
- Rhythm (regular-smooth/firm/flowing-fluent/irregular-rough)
- Timbre (harmonic richness)

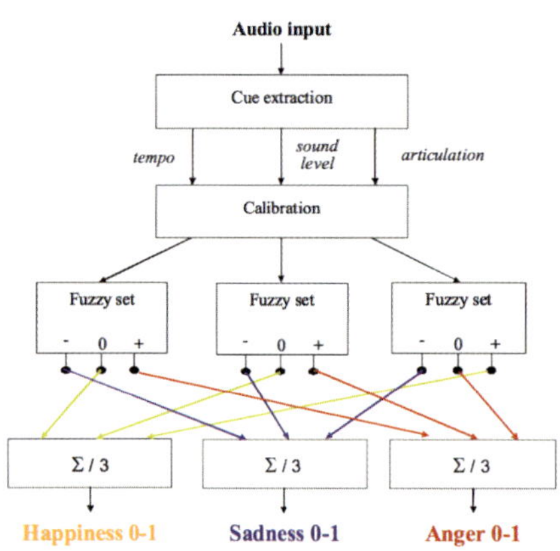

Fig. 13.10. Anders Friberg's fuzzy mapper.

The combination of performance and composer's features define the input in Juslin's *lense model* [54] of emotional performance. These features, alone and interacting with one another, are compared with the output: the emotional judgment of the listener. Using statistical multiple regression analysis, the achievement's quality is measured, yielding a validation of the matching quality between musical structure and emotional effect. Of course, the implicit assumption toward a success of this correlation is that there are adequate features for achieving expected emotions. In this model, the emotional correlates to musical features have been described by Friberg et al. in [8] as follows:

	Happy	Sad	Angry	Tender
Overall Changes				
Tempo	somewhat fast	slow	fast	slow
Sound level	medium	low	high	low
Articulation	staccato	legato	somewhat staccato	legato
Rule				
Phrase arch	small	large	negative	small
Final ritardando	small	-	-	small
Punctuation	large	small	medium	small
Duration contrast	large	negative	large	-

Fig. 13.11. Anders Friberg.

According to this line of thoughts, Friberg has written a program, called "fuzzy mapper" [36], that attributes emotional loadings to musical input (figure 13.10). It takes audio input and extracts three different cues: tempo, sound level (loudness), and articulation. Of course, what they call tempo is not exactly what we would call such, since they are referring to density of sound events in physical time. In a next step, these cues are calibrated and then sent to three units (coined "fuzzy set" in this figure). Each unit calculates three value levels: negative, zero, and positive, for low, middle, and high, respectively. The values are then connected to the emotional semantics boxes, which calculate the averaged sum of these imputs and yield degrees of happiness, sadness, or anger, respectively. For example, the low levels of all three input cues yield (fuzzy values between 0 and 1 of) sadness.

These investigations imply the following results concerning the perception of musically expressed emotions (not induced emotion!):

- Emotional expression can be reliably communicated from performer to listener.
- Up to 80 to 90 percent of the listeners' answers can be predicted using models based on musical features.
- Despite different semantic sets, the four emotions sadness, happiness, anger, and love/tender-ness (including synonyms) seem to be the ones that are especially easy to differentiate, describe, and model.

The association of emotional categories with such basic sound parameters is problematic. Examples abound where the programmed correlations fail. For example, Schumann's first composition in his extremely sad *Gesänge der Frühe* jumps from p to f in the middle, but the dynamics change still amplifies the depressed atmosphere in the moment of this dynamical increase. The program also omits the harmonic and melodic structure, which both are known to contribute essential information about emotional expressivity. We should be aware also that the three components—tempo, sound level, and articulation—are much too coarse to map faithfully onto the huge variety of emotional qualities by the numerical formula as defined in the fuzzy model. Recall that the Circumplex Model discussed in section 13.1 is two-dimensional, and that the reduction of emotional qualities to one number cannot represent two dimensions except if they lie transversally to this numerical reduction.

See also Steven R. Livingstone's Computational Rule System for Modifying Score and Performance in [73] for another computational approach to emotional shaping of performance. He presents a Computational Music Emotion Rule System (CMERS) for the real-time control of musical emotion that modifies features at both the score level and the performance level.

13.6 Alf Gabrielsson's Isomorphism

Fig. 13.12. Alf Gabrielsson.

The previous approach is a thoroughly experimental one and can be tested by statistical methods. It is based upon a correlation of musical features defined by performers and composers with emotional reactions. But it does not claim that (1) every possible emotion can be provoked by adequate musical feature configurations, and (2) every emotion stems from essentially one uniquely determined musical configuration. Although one would consider this as being fairly improbable, there are scholars who believe that music and emotions essentially are in a one-to-one correspondence.

This stream of ideas has been forwarded by philosopher Susan Langer [65]. Here are some of her crucial statements:

- Music represents the dynamic form of emotional life, not specific emotions.
- Music is a tonal analogue to emotive life, music is (...) formulation and representation of emotions, moods, mental tensions, and resolutions.
- Because the forms of human feeling are much more congruent with musical forms than with the forms of language, music can reveal the nature of feelings with a detail and truth that language cannot approach.
- Music has forms of growth and attenuation, flowing and stowing, conflict and resolution, speed, arrest, terrific excitement, calm, or subtle activation and dreamy lapses, patterns of motion and rest, or tension and release, of agreement and disagreement, preparation, fulfillment, excitation, sudden change, etc.

With this philosophical background, the Swedish music psychologist Alf Gabrielsson has set forth a strong hypothesis for the emotional signification of music, namely [42]:

- There is an isomorphism between the structure of music and the structure of feelings (figure 13.13).
- We may also generally notice the fact that we behave/move differently in different moods, as reflected in expressions like "to jump for joy," "sink in despair," "tremble in fear," etc.
- In summary, we may consider emotion, motion, and music as being isomorphic.

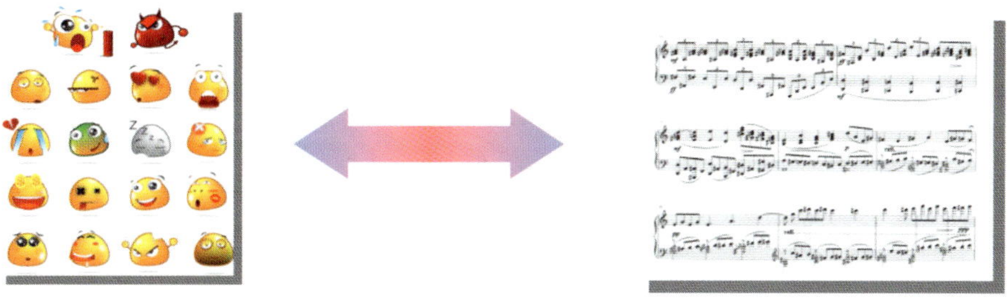

Fig. 13.13. Alf Gabrielsson considers emotion and music as being isomorphic.

Let us comment on these propositions. The first one is stating that two structures are isomorphic. This means that whatever those structures might be, there is a one-to-one correspondence of musical structure and structure of feelings. This llikely means not only that there are as many musical entities as there are feelings, but also that the relations among entities are in such a correspondence with the relations among feelings. This is not explicit, but would make no sense if just reduced to a counting argument. The singular of the wording makes sense only if Gabrielsson means the entire connection between individual musical entities and individual feelings, respectively. We have two isomorphous bodies of human experience. The fact that he uses the word "feelings" instead of "emotions" is not dramatic, since later, in an other context, as shown in the third citation from Gabrielsson, he uses the "adequate" wording.

Let us make this more precise on examples described by Gabrielsson [42]. He connects emotional qualities of music with its structural attributes:

- *Serious and solemn music* is slow, low pitched, and voids irregular rhythms and dissonant harmonies.
- *Sad music* is likewise slow and low pitched and further apt to be in minor mode and to contain dissonance.
- *Calm music* is slow, soft, and apt to contain little dissonance.
- *Happy music* is fast, high pitched, and in major mode and contains little dissonance.
- *Exciting music* is fast and loud and apt to contain dissonance.

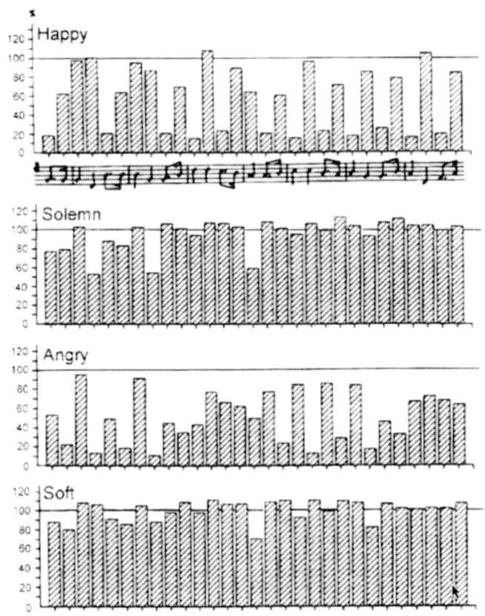

Fig. 13.14. Articulation profiles of *Oh, My Darling Clementine* for different emotional expressions.

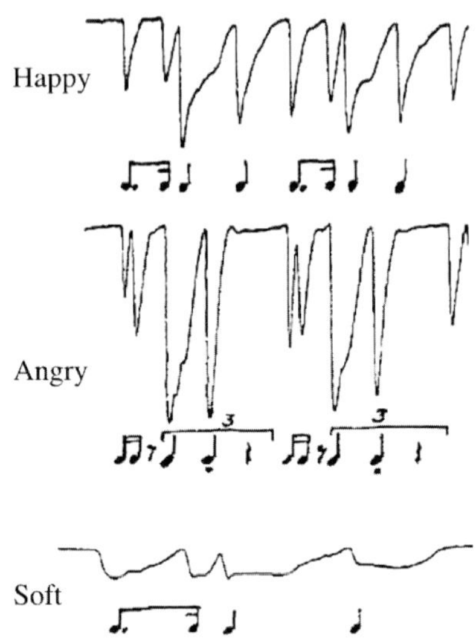

Fig. 13.15. Sentograph pressure profiles of *Oh, My Darling Clementine* for different emotional expressions.

A concrete example is Gabrielsson's measurement of four piano performances of the melody of the song *Oh, My Darling Clementine* (figure 13.14). The quantity measured is the articulation of notes, meaning that 100 percent is nominal duration, and deviations of this duration are given in percent. So a strong staccato would be 20 percent, whereas a legato would be 110 percent. The first performance for an expression of happiness shows strong staccati in contrast to intermittent weak staccati and some rare legati. Anger shows a very similar image, however less strong staccati and no legati, and the alternation between strong and weak staccati is less prominent. While these characteristics may be valid for a distinction between happiness and angriness in this piece, the distinction between solemness and softness is far less characteristic. Figure 13.15 shows a second measurement, this time with Clynes' sentograph (see next chapter), measuring the finger pressure according to three different emotions.

More generally speaking, the characterization of calm music by a small number of dissonances is not universal. For example, if we look at the lyrical interpretation of standards, such as *When I fall in Love* by pianist Bill Evans in [28], we recognize a huge number of dissonances, while the piece is slow and soft and expresses a beautiful calmness.

There are more serious problems with Gabrielsson's claim. To begin with, his presentation is far from what would be understood as being an isomorphism. There are some correspondences, but they are not one-to-one even in the given shape, as shown above. And they are far from constructive and a fortiori far from systematic. We would need a number of explicit principles in order to construe such an isomorphism. How would it be connected to the inner narrative (the juncture of plan) of an emotion, according to Lazarus' categorization, or to the metrical topology as proposed by Russell's Circumplex Model?

Besides the epistemological deficiency of this claim, we also have to consider the structure of music and guess whether such an isomorphism is likely to exist given the complexity of this structure. Let us, for example, look at all isomorphism classes of melodies of 72 tones, given in onset and pitch (not even considering differentiation in loudness, duration, or instrumental color), and identified according to standard criteria of affine geometry (transposition, time shift, inversion, retrograde, arpeggio, fifth circle, dilation, see [84, Chapter 11] for details), then we obtain

$$2\ 230\ 741\ 522\ 540\ 743\ 033\ 415\ 296\ 821\ 609\ 381\ 912 \sim 2.23 \times 10^{36}$$

isomorphism classes, which are, compared to the roughly 10^{11} stars in a galaxy, near to infinity in terms of structural richness. So how could Gabrielsson's isomorphism be compatible with this richness? There are two solutions: Either this classification is musically irrelevant, or the richness of emotions must be updated to cope with this motivic repertory. In the first case, we have to understand how musical structures must be described, grouped, and classified to enable such an isomorphism. But even if that worked, the reduction of such classification would mean that the very concept of a musical structure must be rethought and must also be capable of pushing back the present analysis of musical structures to some field of musically irrelevant mathematical ordering scheme. In the second case, we would have to delve into a thorough investigation of the universe of emotions that would eventually give us the required richness. From the given approaches, I see only the Circumplex Model as being capable of producing such a variety, namely by continuous variation of coordinates of points in the model's plane. However, one would then have to ensure the human capability to distinguish the coordinate values in pleasantness and activation in such a fine granularity that would add up to the scale of the above number. I guess that either of these alternatives is problematic, to say the least.

Despite these caveats, the program of searching for relations between these two domains remains a very important one in terms of emotional semantics of music and its expressivity.

14

Gestural Expression

Dans le vide de l'espace quelqu'un dessine,
Crée à travers son corps l'infini du temps.
Marcel Marceau

Gestural expression is a difficult topic, not because the concept of a gesture is so difficult, but because it looks so obvious that everybody believes to know what it means. This first impression is, however, misleading and it is there that the difficulty arises. The word "gesture" is like "time": If your are not asked, what it means, you know, but if you are asked you cannot tell. Saint Augustin's famously articulated this fact when asked about time. We shall not give a precise definition of a gesture until section 14.8. Instead, we want to work with an intuitive understanding.

14.1 General Facts About Gesture Theory in Music

Despite the consciousness that gestures play an important role in music, there has not been such a thing like a gesture theory in music. This was made explicit by music theorist Robert S. Hatten in his book *Interpreting Musial Gestures, Topics, and Tropes* [47]: "Given the importance of gesture to interpretation, why do we not have a comprehensive theory of gesture in music?" Gesturality became an important topic when Hatten realized that performance of classical piano music—Mozart, Beethoven, Schubert—is strongly determined by gestural utterance. Although Hatten's investigations are focusing on gesture, they are not formalized but remarkably subtle. His definition of a gesture reads as follows: "Gesture is most generally defined as communicative (whether intended or not), expressive, energetic shaping through time (including characteristic features of musicality such as beat, rhythm, timing of exchanges,

G. Mazzola, *Musical Performance*, Computational Music Science,
DOI 10.1007/978-3-642-11838-8_14, © Springer-Verlag Berlin Heidelberg 2011

contour, intensity), regardless of medium (channel) or sensory-motor source (intermodal or cross-modal)."

On a more formal level of music theory, David Lewin, the *monstre sacreé* of American music theory, has indicated that lack of gesturality in his seminal book *Generalized Musical Intervals* [68]. Lewin suggests that transformations between musical points (pitch classes, for example) are the new path to pursue. In [68, p. 159], we read: "If I am at s and wish to get to t, what characteristic gesture should I perform in order to arrive there?" Although his transformational theory is in fact not a gestural one yet (it is merely the step into mathematical—more precisely, category-theoretical—process diagrams), Lewin thinks in a gestural spirit and repeats this attitude in another statement, still relating to his question about the movement of s to t. He adds [68, p. 159]: "This attitude is by and large the attitude of someone *inside* the music, as idealized dancer and/or singer. No external observer (analyst, listener) is needed."

Interestingly, Hatten's and Lewin's approaches are based upon a quite blurred concept of a gesture, but they have the merit to have made that important point. It is not only in academic music theory that this insight has been forwarded. Piano giant and music philosopher Cecil Taylor also pointed out, "The body is in no way supposed to get involved in Western music. I try to imitate on the piano the leaps in space a dancer makes."

So the gestural dimension is recognized, but not elaborate, on a theoretical level. This has strong impliciations for performance theory of gestures: Everybody agrees that performance of gestural contents and performance via rhetorical application of gestures is crucial, but when it comes to the explicit, detailed discourse, the topic evaporates like that beautiful, but equally ghostly, discourse about playing the body of time. Although a number of gesture-driven investigations and implementations have been realized by computer music communities [146], these contributions do not improve the conceptual control of the topic because they are restricted to applying the physical gestures as spatial movements of the body's limbs to specialized interfaces with electronic musical instruments.

14.2 Roger Sessions: Gestures in Performance

The dramatically intense but still underestimated role of gestures in performance has been described in a beautifully clear way by American composer and music critic Roger Sessions in his book *Questions About Music* [125, Chapter III]:

> *It is fairly obvious, I suppose, that our total awareness of movement—which in essence signifies our awareness of* time as a process*—demands sustained attention, which is limited to the duration of the specific act*

of movement in question; it holds us captive, as it were, for the duration. We are aware of a beginning and an end. In respect to space on the other hand, the words "beginning" and "end" have an essentially metaphorical meaning; they represent boundaries or limits that remain even after we have become aware of them, as does all that lies between. Our attention is our own to husband and deploy as we wish. We can withdraw it and absent ourselves merely by averting or closing our eyes, and return whenever and for as long as we wish.

What I am saying is that we experience music as a pattern of movement, as a gesture; and that a gesture gradually loses its meaning for us insofar as we become aware of having witnessed it, in its total identity, before. If it is to retain this meaning in its full force, it must be on each occasion reinvested with fresh energy. Otherwise we experience it, to an increasing degree, as static; its impact, as movement, diminishes, and in the end we cease to experience it as movement at all. Its essentially static nature has imposed itself on our awareness.

This is why I am convinced that the performer is an essential element in the whole musical picture. It is why I came to realize that my earlier dreams—that composers might learn to freeze their own performance, in wax or otherwise (tape recorders had not been invented at that time)—were, to put it bluntly, quite ill-directed. They were ill-directed, above all, for the reasons I have been outlining; a gesture needs constant renewal if it is to retain its force on subsequent repetitions. Composers above all should know this, especially if they have developed the practice of taking part in performances of thir own work. Each performance is a new one, and the work is always studied and approached anew, even by the composer. The same, it should be obvious, is true of professional performers. I would go even much further and point out that there is no such thing as a "definitive" performance of any work whatever. This is true even of performances by the composer himself, in spite of the fact that recordings of his performances of his own work should be made and preserved, for a number of quite obvious reasons.

Session's discussion of movement as a processing of time leads him to acknowledge that this dynamical action is a gesture—not only in the making, but also in the music's perception. So he gives the argument for a messaging of gestures, and by means of gestures, which is our topic in this chapter. It is remarkable that he then recognizes that a gesture cannot preserve its meaning except in its energetic refreshment on each occasion of performance. This is very similar to the French theory of gestures [91, Chapter 7.2], which stresses the impossibility to tame living gestures.

He moreover recognizes the performer's essential role in the "whole music picture" and also reminds composers, himself included, that their work of musical creation is not accomplished until it is performed. This does not mean that a composer must intervene in the performance of his/her works. Some are dead and simply cannot do this anymore. No, it means that the completion of a musical work cannot be achieved before its performance has occured. In this sense, performance is strongly what semioticians call a *deictic* part of the musical sign system: Musical signs reach their full meaning only and essentially through their pragmatic instantiation.

Fig. 14.1. Roger Sessions.

This second insight is strongly related to the gestural aspect since gestures are not lexicographic, they are shifters, as Sessions stresses with his "French" view on gestures. We are not astonished that Manfred Clynes refers to Session's writings in his critique of score-based music. We come back to the gestural aspect in Clynes' work later in section 14.5.

14.3 Theodor Wiesengrund Adorno's Gesture Theory in Musical Reproduction

Theodor W. Adorno has written deep analyses of performance, in particular with respect to their subcutaneous gestural implications in his posthumous work *Zu einer Theorie der musikalischen Reproduktion* [2]. It is interesting to see how Adorno gets off the ground with his gestural discourse on the same basis as Sessions and Clynes, namely a radical critique of the score-based reduction of music (translated from [2, p.227/8]):

> *Notation wants music to be forgotten, in order to fix it and to cast it into identical reproduction, namely the objectivation of the gesture, which for all music of barbarian cultures martyrs the eardrum of the listener. The eternization of music through notation contains a deadly moment: what it captures becomes irrevocable.*
>
> (...)
>
> *Spatialization (through notation) means total control. This is the utopic contradiction in the reproduction of music: to re-create by total control what had been irrevocably lost.*
>
> (...)
>
> *All making music is a recherche du temps perdu.*

And later on (translated from [2, p.235]):

> *Musical notation is an expression of the Christianization of music.*
>
> (...)
>
> *It is about eternity: it kills music as a natural phenomenon in order to conserve (or "embalm" G.M.) it—once it is broken—as a spiritual entity: The survival of music in its persistence presupposes the killing*

> *of its here and now, and achieves within the notation the ban (or "detachment" G.M.) from its mimetic*[1] *representation.*

To begin with, Adorno, Sessions, and Clynes agree upon the fact that music notation, and its score, abolishes music, which is fixed and cast into a format for identical reproduction. It does so in objectifying the gesture and thereby martyring the eardrum, an act of barbarian culture. It is remarkable that musical notation is related to barbarian culture. The eternization of music in the notation's casting is killing music; it retains a dead body, not the living music. This eternity of dead—in fact, embalmed—bodies appears as a Christian ritual of sacred denaturation. The procedure of notation kills the music's here and now; its expressivity is annihilated, banned forever. The notational process kills through spatialization, which means total control, time does not fly by anymore, a note is a point in a dead space of eternity. Adorno views this as being the great contradiction of notation in that it claims total control for a reproduction of what has been irrevocably lost. It is a *temps perdu*, and making music is doomed to a *recherche du temps perdu*.

Adorno then makes important comments on what he views as being the gestural substance of music (translated from [2, p.244/5]):

> *As each face and each gesture, each play of features, is mediated by the I, so the musical moments are the very arena of mimic in music. What must be read and decoded within music are its mimic innervations.*
>
> (...)
>
> *However, a pathetic or cautious or expiring location does not signify pathos, caution, or expiration as a spiritual thing, but maps the corresponding expressive categories into the musical configuration, and those who want to perform them correctly have to find those encapsulated gestures in order to mimic them.*
>
> (...)
>
> *Finding through reading: the decoding work by the interpreter; the very concept of musical performance is the path into the empire of mimic characters.*
>
> (...)
>
> *The spatialization of gestures, that impulse of neumatic notation is at the same time the negation of the gestural element.*
>
> (...)
>
> *By the visual fixation, where the musical gesture is positioned into a simultaneous relation to its equals, it ceases to be a gesture, it becomes an object, a mental thing.*

Here Adorno refers to the mimetic category in his theory. It is the category "expression of expression." So it is about the expression of emotions,

[1] For Adorno, "mimesis" means "expression of expression," and this is precisely our context: The expression as content is expressed via rhetorical shaping.

for example, not about emotions, and it is about the musical image of these expressions. Therefore, we have to read those mimic innervations of gestural expressions in music. Musical performance deals with the explication of those hidden innervations, with the action of displaying them in the making, here and now. And it now becomes clear that the neumatic notation creates static photographs of those gestures, which negate them by this spatial fixation. The spatial trace of a gesture is its negation, freezing it as a spatial object.

We should, however, briefly digress on the very concept of a space here, since it is not what a geometer or a physicist would call a space. In physics, a space is a geometric entity that can have different interpretations, so space-time is (locally speaking) a four-dimensional real vector space, and the mathematical structure of time is not different from that of the three space coordinates. Of course, the Lorentz metric distinguishes time in the metrical structure of space-time, but it is still a metrical space. In performance theory of music, time has a radically different role. The four-dimensional space of onset, pitch, loudness, and duration for piano music, which is used in score notation, does not have the ontology of musical time. Under no circumstances would the onset or duration coordinates be accepted as representing the time that takes place in performance. This *differentia specifica* in the performative time concept is related to gestures, not to geometric representation. For Adorno, gesture has an existential character; it cannot be objectivized; it only exists in the moment of the making; it is mediated by the I, which cannot be cast in a dictionary—the I is the non-lexical, the shifter, par excellence. However, it is not part of the subject, it is not subjective as opposed to being objective (the score objects are so). I is only mediated by the I, it seems to lie between subject and object; therefore, the utterance of a gesture is neither object nor subject.

Adorno continues (translated from [2, p.269]):

> *The true reproduction is the mimicry of a non-existent original.*
> (...)
> *But this mimicry of the non-existent original is at the same time nothing else but the X-ray photograpy of the text.*
> (...)
> *Its challenge is to make evident all relations, transitions, contrasts, tension and relaxation fields, and whatever there is that builds the construction, all of that being hidden under the mensural notation and the sensorial surface of sounds.*

The true reproduction is not a reference to an object out there; the original is non-existent, and it is not the I, which would be an existent entity. It is something mysterious since there is an X-ray procedure, but it does not show something hidden in the dead object of the score. It is as if that mystery would be brought to existence by the very X-ray procedure. The innervation must be made, not only discovered and pointed to.

Adorno's concept of a gesture is as difficult as it is radically different from what can be described in terms of traditional subject-object duality.

Let us see what Adorno concludes from all these subtle reflections (translated from [2, p.269,270,271]):

> *What happens in true performance is the articulation of the sensorial appearance that reaches into the most hidden details, wherein the totality of the construction, the gesture of the work, reveals its mimical execution.*
>
> (...)
>
> *The concept of clarity defines the degree of an analytical performance: everything that exists as relations within the mensural text must become clear, but this concept cannot be understood in a primitive way, i.e. as a clarity of every single relation, but as a hierarchy of clarity and blurredness in the sense of the clarity of the overall structure, the mimic gesture.*

And he summarizes this entire perspective on gestural performance (translated from [2, p.247]):

> *Correspondingly the task of the interpreter would be to consider the notes until they are transformed into original manuscripts under the insistent eye of the observer; however not as images of the author's emotion—they are also such, but only accidentally—but as the seismographic curves, which the body has left to the music in its gestural vibrations.*

14.4 Renate Wieland's Gestural Piano Pedagogy

As a student of Adorno, piano pedagogue Renate Wieland, in collaboration with her colleague Jürgen Uhde, has developed a theory of piano performance that is based upon Adorno's gestural philosophy. The remarkable feature of this work is that she succeeds in

(1) giving her approach a clear-cut separation from emotional dramaturgy and

(2) reshaping gesture theory in an explicit geometric language.

Fig. 14.2. Renate Wieland.

She makes these two points very clear in her text (translated from [140, p.169]):

> *Musical gestures are perceived in the free conducting movement, in the playing movement and sublimated in the spiritual mimesis of pure imagination. Whatever the level, such experiments are always within space. Originally, affects were actions, related to an exterior object,*

> *along the process of interiorization they were detached from their object, but they are still determined by the coordinates of space.*
>
> (...)
>
> *Language reminds us everywhere of the connection of affect and movement and of the way gestures behave in space. It speaks about hautiness, elevation and inclination, about greatness of mind, pettiness, about respectful and forward, etc.*
>
> (...)
>
> *There is therefore something like gestural coordinates; they can help ask how the gestural impulse out of the inner is projected into space, how it wants to expand, which direction is dominant: Is its energy vertically or horizontally active? Does it rather propagate ahead or backward? Upward or downward? To the right or to the left? Are forces acting more concentrically or excentrically? Does the gesture rather point "inward," as we read in Schumann's work, or "outward"? Which amplitude does the expression choose? Does it live in all spatial dimensions, and with what proportion and intensity?*

She reminds us of the etymology of the word "emotion": *ex movere*, to move from inside out. She also makes clear that the original setup is now internalized, but that it remains a spatial concept. She then gives examples of etymological shifts, which are parallel to this internalization process: Words now mean abstract things, but when we go to the kernel of that meaning, it is related to a spatial action. So the mimetic action in Adorno's sense is the expression of that spatially conceived gesture in the realm of musical space.

She adds the following excellent illustration of a gestural mimesis in music (translated from [140, p.169]):

> *Models of contrast between extreme vertical and horizontal gestures are found in Beethoven's Bagatelle op.126,2.*
>
> (...)
>
> *Aggressively starting initial gestures are answered by flat, conciliating gestures, where the extremes are polarized to the outermost in the course of the piece. In this way, asking again and again, gesture becomes plastic in the end. But it only succeeds insofar as it constitutes a unity, is emanated from one inner central impulse.*
>
> (...)
>
> *Gestures are the utmost delicate; where their unity is disturbed, their expression immediately vanishes.*

It is again in Adorno's and Session's spirit that she views gestures as being extremely unstable in their existentiality: Nothing is easier than to disturb and vaporize a gesture. It is by this fact that Gilles Châtelet, one of the fathers of French gesture theory, has characterized gestures as being the smile of existence [14].

Wieland finally transcends her approach in a seemingly breathtaking intensification, which reads as follows (translated from [140, p.190]):

The touch of sound is the target of the comprising gesture; the touch is so-to-speak the gesture within the gesture, and like the gesture at large, it equally relates to the coordinates of space.

(...)

The eros of the pianist's touch is not limited to the direct contact with the key, the inner surface of the entire hand pre-senses the sound. etc. etc.

She introduces what one could call the reverberance of a gesture, namely the gesture within a gesture, meaning that a gesture can incorporate other gestures, can become a gesture of gestures. We shall see later in section 14.8, relating to our own research, that this concept is very powerful for the theory of gestures in that it enables complex imbrications of gestures, so-called hypergestures, for the construction of movements of movements of movements..., an idea that is crucial in the dynamics of musical utterances.

14.5 Manfred Clynes' Essentics as a Theory of Gestural Expressivity

Fig. 14.3. Tamas Ungvary playing the sentograph. The joystick is accessed with the right middle finger.

We saw in section 13.3 that Manred Clynes conceived expressivity as a shaping of performance in pulses, those embodiments of essentic forms, via specific deformations of duration and loudness. He claimed that such pulses were characteristic for the emotional expressivity of composers such as Beethoven, Mozart, etc. Clynes' pulses are not only emotional categories, but also, and perhaps more significantly, gestural utterances. The curves associated with pulses, as shown in figure 13.7, are gestural shapes, movements in the space of duration and loudness. Clynes accordingly constructed and patented a machine, the *sentograph*, providing us with an interface to grasp such gestural movements. Following Clynes' ideas, Hungarian composer Tamas Ungvary has constructed a sentograph that can be used by improvising composers in order to play/create music by gestural input [141]. Ungvary replaces the usual encoding of sound events in discrete points in a parameter space by an intrinsically gestural input that is given by variable pressure and angle on a joystick (figure 14.3). Despite the fascinating perspective on musical creation, the gestural input remains very abstract insofar as no significant movement of the fingers is possible. The musician has to stay in contact with that fixed piece of metal and cannot move freely in space. This restriction heavily limits the natural human need for movements when gestures have to be created from the living body. Perhaps a more natural encoding of the input parameters would improve the expressive power of this interesting machine.

14.6 Johan Sundberg, Neil P. McAgnus Todd: Mechanical Models of Gestures in Music

On a more down-to-earth level, gesture has been studied by Johan Sundberg and collaborators. In a paper entitled "Is the Musical Ritard an Allusion to Physical Motion" [64], Sunberg and Ulf Kronman have studied final ritard as a phenomenon akin to physical ritard. The model conjectures that a tempo decrease at the end of a musical piece would be related to a quadratic function, which appears for mechanical ritard with a constant force. So we are given a constant force F, and its action on a given mass m, which generates a constant deceleration $a = F/m$ according to Newton's second law. Given an initial velocity v, the velocity after t seconds is $v - a.t$. Whence the distance $s(t)$ traveled after t seconds is $s(t) = \int_0^t v - a.\tau d\tau = t.v - a/2.t^2$. If the final velocity at time t_0 is 0, we have $t_0.a = v$, whence $s(t_0) = (v/a).v - a/2.(v/a)^2 = v^2/2a$. Therefore velocity at time t is $v(t) = v.\sqrt{1 - s(t)/s(t_0)}$. Supposing that this physical situation relates to the musical one by a constant c, i.e. $s(t) = c.E(t)$, E being the symbolic onset, we get $T(t) = T(t_0).\sqrt{1 - E(t)/E(t_0)}$. In other words,

$$T(E) = T(E_0).\sqrt{1 - E/E_0}$$

namely the tempo at onset E being the above function of the tempo $T(E_0)$ at the beginning E_0 of the ritard, the onset E and the beginning onset E_0. The experimental situation is shown in figure 14.4. The parabolic tempo curve relates to the phase I in the left graphic. Phase II is interpreted as a linear tempo decrease.

Besides the poor fit of the measured tempo with the mathematical curve, the question arises why such a mechanical function should hold. What is the musical analog to mass, what is the force analog to a constant mechanical force? We do not see any musical structure entailing such a mechanical model. It is interesting that the ritard phase II relates to a quite sophisticated harmonic and melodic musical process, which is not taken into account.

Another mechanical model of agogics has been proposed by Neil P. McAgnus Todd in [137]. He rightly observes that the final retard is only a very special agogical situation and therefore models his tempo curves according to a superposition of accelerando/ritardando units that are defined by a triangular sink potential V. Accordingly, tempo is defined as a velocity v, and the total energy of the system $E = \frac{1}{2}mv^2 + V$—supposed to be constant (why so?)—gives the velocity formula $v = \sqrt{2(E - V)/m}$. Todd further supposes that there is an intensity variable I for loudness, with a relation $I = K.v^2$ that is common to many physical systems. This yields the relation $I = 2K(E - V)/m$ and sums up to an aggregated formula $I = \sum_l 2K(E - V_l)/m_l$ if the grouping

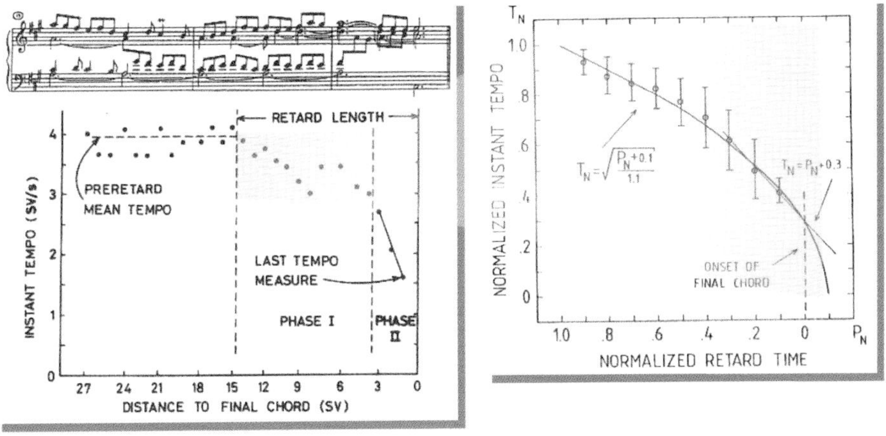

Fig. 14.4. The parabolic tempo curve (right figure) relates to the phase I in the left graphic. Phase II is interpreted as a linear tempo decrease.

of the piece is taken into account. The idea is that there is a physical energy and intensity parameter system that controls the "surface" of the tempo (= velocity) via classical energy formulas. The background structure is an energetic one, i.e., the tempo curve and loudness are expressions of mechanical dynamics. The author comments on his method as follows [137, p.3549]:

> *The model of musical dynamics presented in this paper was based on two basic principles. First, that musical expression has its origins in simple motor actions and that the performance and perception of tempo/musical dynamics is based on an internal sense of motion. Second, that this internal movement is organized in a hierarchical manner corresponding to how the grouping of phrase structure is organized in the performer's memory.*

The author also suggests a physiological correlate of this model (loc. cit.):

> *...it may be the case that expressive sounds can induce a percept of self-motion in the listener and that the internal sense of motion referred to above may have its origin in the central vestibular system. Thus, according to this theory, the reason why expression based on the equation of elementary mechanics sounds natural is that the vestibular system evolved to deal with precisely these kinds of motions.*

Todd refers to the insights of neurophysiologists that the vestibular system is also sensitive to vibrational phenomena. The musical expressivity is therefore understood as an effect of transformed neurophysiological motion.

The drawback of this approach is that finer musical structures are not involved in the structuring of the energy that shapes tempo/intensity. And even if that could be done, there is an essential kernel of this shaping method that should be based upon paradigms of motion. These paradigms do not however appear clearly in the above approach. More precisely: The complex motion dynamics of the vestibular system cannot easily be mapped onto the structures of performative expressivity. What is the operator that transforms whatever structures of motion into expression parameters? If music were isomorphic to motion, no such isomorphism could be recognized from Todd's approach.

Fig. 14.5. The gesturality of four famous pianists: Vladimir Horowitz (top left), Glenn Gould (top right), Arturo Benedetti Michelangeli (bottom left), Cecil Taylor (bottom right)

14.7 Guerino Mazzola and Stefan Müller: Modeling the Pianist's Hand

One of the most evident gestural expressions is the body movement of a performing musician. We instantly recognize the gestural power of four famous pianists as shown in figure 14.5. It is logical that one should therefore attempt to model gestures of musicians. In collaboration with my PhD student Stefan Müller, we embarked in the modeling of the pianist's hand [85] on the level of computer graphics. The idea was not only to model the hand's movements, but also to implement a software that could transform the abstract symbols of a score into hand movements that were adequate for the rendering of the score on a piano keyboard.

The project had three components:

1. Modeling the hand with its spatio-temporal trajectory in the movement.
2. Transforming abstract score symbols of notes (what we call deep-frozen gestures, since they historically stem from neumatic abstraction, neumes being gestural signs) into *symbolic gestures*, i.e. curves in a space related to the piano keyboard geometry.
3. Deforming symbolic hand gestures into physically valid spatio-temporal curves of the pianist's hand.

14.7.1 Modeling the Hand

This task was accomplished with a simplified representation of the hand by six curves $\gamma_i(t)$ in physical space-time with space axes x, y, z to denote the momentous position of the hand, and e, the physical time of that position. The curve parameter t is not the physical time, it is just an abstract curve parameter (figure 14.6). The curves $\gamma_1, \ldots \gamma_5$ represent thumb, index, middle, ring, and little finger, respectively, while γ_6 represents the carpus.

These curves are subjected to geometric constraints G resulting from their connectivity as parts of the hand's geometry. We refer to [85] for more details. And the curves are subjected to mechanical constraints M, which means that if we think of the ith finger's mass m_i, and if the pianist is capable of exerting a maximal force of K_i upon that mass, then Newton's second law imposes the inequality

$$m_i|d^2\gamma_i^{space}/de^2(t)| \leq K_i$$

at any curve parameter value t, where γ^{space} is the three-dimensional spatial part of the curve.

14.7.2 Transforming Abstract Note Symbols into Symbolic Gestures

Refer to figure 14.7 for the following discussion. In traditional performance theory, we look at the transformation $\wp_{score}$ of score symbols into sound events. This is shown in the bottom row of the rectangular diagram of figure 14.7. In the gestural extension of this disembodied process, we have to create the sonic result via gestural actions. The sounds are just the result of physical gesture curves interacting with the keyboard; these curves are shown in the right top corner of the diagram.

In order to generate these physical curves, one first has to unfreeze the note symbols and to transform them into gestural symbols. This unfreezing process is shown in the left top-bottom half of the diagram. This does not create physical gesture curves, but only symbolic curves, which are faithful

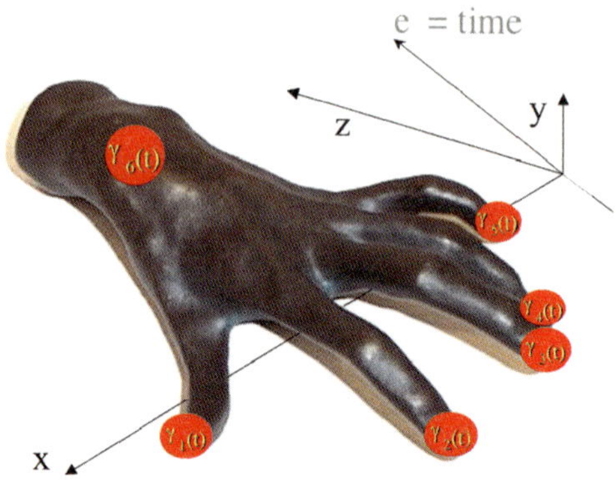

Fig. 14.6. Modeling the pianist's hand.

representations of the note symbols. This process resembles the MIDI interpretation of notes insofar as the commands associated with notes are abstract movements: In MIDI, a note is defined by an ON command, which means go down to that pitch at a determined moment, and the MIDI velocity used to move down to the key defines loudness. Then the finger remains in that position until the OFF command tells it to move up again, etc.

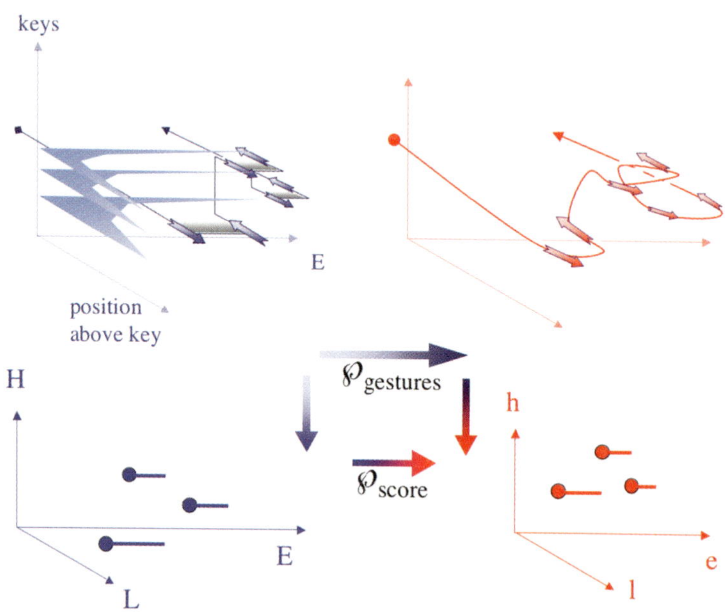

Fig. 14.7. The four levels of performance: symbolic score representation (left bottom), performed sound events (right bottom), symbolic gesture curves (left top), physical gestures (right top).

This representation defines a very abstract curve, but it is this that tells the fingers in a qualitative way how to move. This movement is shown in the left top corner. We see the symbolic gesture associated with a sequence of three notes in the left bottom corner. The finger moves down over a first key, then remains there and after the duration moves up, changes the key coordinate, goes down to the second key with a second velocity, remains there for its duration, moves up, shifts to a third key position, moves down with the third velocity, remains for that duration, and finally moves up. All these phases are connected in a continuous curve, which has angles, i.e., it is not differentiable, and whose movement is orthogonal to the time axis E when moving down at a determined velocity.

14.7.3 Deforming Symbolic Hand Gestures into Physically Valid Curves

The third step toward gestural performance is the horizontal transformation on top of the diagram in figure 14.7. The given symbolic curve does the right thing, but it does not move within the geometric and physical constraints. These constraints define a subspace of the space of all continuous curves, in fact a manifold $X(G, M)$ in terms of global geometry. We are given the symbolic gesture curve from the left top data and now have to create a pysically valid deformation thereof, i.e. one that fulfills the geometric and mechanical constraints G, M. This is a very delicate operation. Essentially, it boils down to looking at the symbolic curve $\gamma_{\mathrm{Symbolic}}(t)$ and then searching for one $\gamma_{\mathrm{Physical}}(t)$ that is as near as possible to $\gamma_{\mathrm{Symbolic}}(t)$ and lives in $X(G, M)$. The delicate point is that it is often not possible to cope with all conditions for a perfect performance, since, for example, physics does not allow for infinite velocities. So when the finger has to play two different keys in immediate succession without a pause, the duration of the first note must be shortened in order to jump from the first to the second key. Such difficulties must be met by defining distance between curves in such a way that musical constraints are given a high weight. For example, the prescribed key coordinates cannot be changed, while durations may be changed, but only a little, etc. It may then happen that there is no solution to a given score input and its associated symbolic curve. This must be possible as a function of the anatomic and physical constraints given from the human conditions. We have implemented this process and have performed a simple Czerny exercise (the same as was used for the Espresso Rubette shown in figure 10.4 in section 10.3), illustrated in figure 14.8.

14.8 A Mathematical Gesture Theory

The desire of a precise definition of a gesture was already in the air following all those more or less philosophical and aethetical approaches to gestures, which we have discussed previously. Following the experimental implementation of a

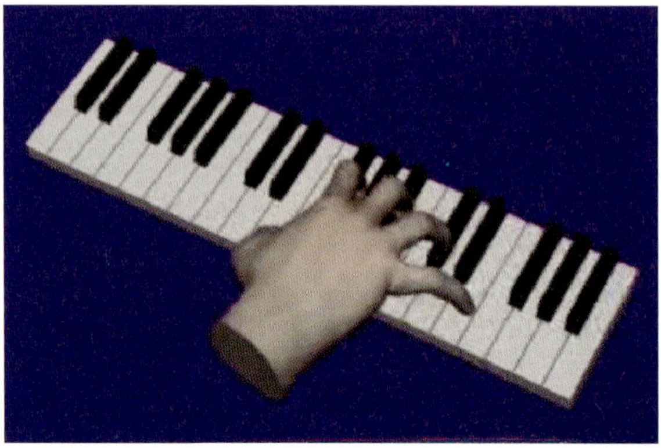

Fig. 14.8. A Czerny exercise played by the computer-graphical model of a pianist's hand according to the gestural "unfreezing" process described in the text.

performing pianist's hand in [85], a more explicit definition of a gesture became feasible. We also discovered that there was a very interesting definition around that came from medieval studies. Medieval theologist Hugues de Saint Victor defines: "Gestus est motus et figuratio membrorum corporis, ad omnem agendi et habendi modum." Gesture is the movement and figuration of the body's limbs with an aim, but also according to the measure and modality proper to the achievement of all action and attitude. Most important is that it is an articulated figuration, a composition of parts (limbs), and that it includes a movement of that figuration in the space-time of the given body. Moreover, it serves for any (omnem) mode of action and attitude, so it has a purpose or target, but it does not, automatically, point to a semantic level—it only reaches the mode of an activity/habit. So this concept follows Adorno's and even more strongly Wieland's gesture philosophy, which is a spatial one, without including semantics automatically; it is only "expression of expression."

Following this approach, we define a gesture as being (a) a directed graph $\mathcal{D}$, called the gesture's *skeleton*. This is the schematic description of the configuration of limbs in Saint Victor's definition. Then we need (b) a map g that associates with each arrow a of $\mathcal{D}$ a continuous curve $g(a) : I \to X$ defined on the unit interval $I = [0, 1]$ of the real number line in such a way that matching arrows carry over to matching continuous curves. The system of these continuous curves is called the gesture's *body* (figure 14.9). It is in this latter part of the gesture that movement of the gestural configuration takes place—for example, if the space X is a space-time, i.e. contains time as a coordinate, such as we had defined gestural curves in the modeling of the pianist's hand in section 14.7.1. The space of all continuous curves in X is denoted by $\overrightarrow{X}$, meaning that it is a (big) directed graph, whose arrows are the continuous curves, and whose vertexes are the points of X. Then the gesture can then be written as a map of directed graphs $g : \mathcal{D} \to \overrightarrow{X}$.

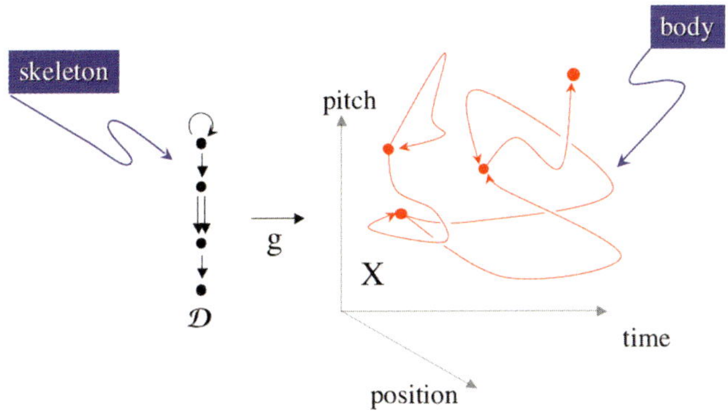

Fig. 14.9. A gesture is a map g from the digraph $\mathcal{D}$ (the skeleton) into the system of continuous curves in a topological space X, the gesture's body. In this example, X is the space with coordinates used for the fingertip positions of a pianist's hand at a given key (pitch), a level above the key (position), and the time of this event.

Two gestures $g : \mathcal{D} \to \overrightarrow{X}$ and $h : \mathcal{E} \to \overrightarrow{Y}$ can be related to each other by a kind of function called *morphism of gestures*, which is a pair $u : \mathcal{D} \to \mathcal{E}, v : X \to Y$ with u being a map of directed graphs, and v being continuous, such that $h \circ u = \overrightarrow{v} \circ g$; see figure 14.10 for an example of such a morphism that connects gestures in multitouch spaces to gestures in musical spaces. This situation is typical in gesture theory: We have gestures in different topological spaces and need to connect them via auxiliary maps between their skeleta and/or between the topological "carrier" spaces.

The most powerful device in this mathematical theory of gestures is the concept of a hypergesture. Recall from section 14.4 that Renate Wieland mentioned that mysterious idea of a "the gesture within the gesture." This construction is now fairly easy: It can be shown that the set of gestures from a fixed skeleton $\mathcal{D}$ to a fixed topological space X is itself a topological space, denoted by $\mathcal{D}\overrightarrow{@}X$. Therefore we may consider gestures $h : \mathcal{F} \to \mathcal{D}\overrightarrow{@}X$. Such gestures are called *hypergestures*: gestures whose body is a system of curves of gestures! Although this sounds complicated, it is quite intuitive. For example, if we have a gesture with a loop as a skeleton, and then a hypergesture with again the loop as skeleton. Then the hypergesture is a loop of loops, in fact a nice closed tube, as in figure 14.11.

Fig. 14.11. A closed tube is a hypergesture, namely a loop of loops.

Such hypergestures are very useful in generating gestural interpretations of classical musical compositions. For example, Beethoven's tonal modulation $B_\flat major \to Gmajor$ in the beginning of op.106, Allegro, can be described by use of such hypergestures, see [93].

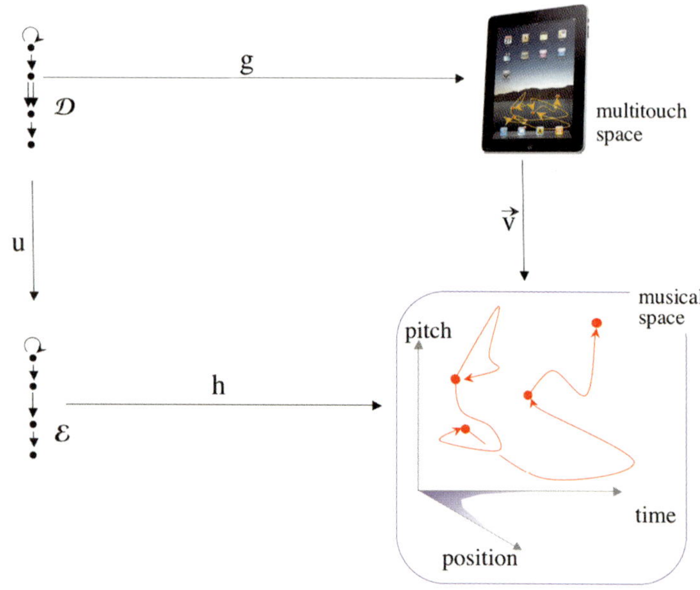

Fig. 14.10. Gestures g and h may also be connected to each other by a so-called *gesture morphism*, which means that the digraphs and the topological spaces are provided with morphisms, respectively, that are compatible with the two gestures (commutative diagrams).

14.9 Anders Friberg, Antoni Camurri et al.: Computer-aided Emotional Analysis of Performance

We terminate our tour d'horizon of gestural expressivity with a short review of recent attempts to incorporate gesture and emotion into the analysis of performance.

In research related to a EU program (MEGA) [36], [12], Anders Friberg, Antonio Camurri and collaborators implement a software that analyzes emotions from music and body gestures and then generates an emotional valuation thereof, yielding a fuzzy mixture of happiness, sadness, and anger. The system takes cues from music tempo, dynamics, and articulation, from body motion, and from overall motion (via video displacement pixels), and valuates the motion and music performance cues (large, fast, uneven, jerky for motion/anger, etc.; loud, fast, staccato, sharp timbre for musical anger, etc.). Then after a gauging/standardization step, the three fuzzy values (0 to 1) of the three emotions are calculated (by the fuzzy mapper, discussed in section 13.5, see also figure 13.10). The expression mapper can be used to visualize the emotions by color and shape (see figure 14.15 for its process flow chart). This visualization was implemented by Camurri and is shown in figure 14.12. The gestural processing of Camurri's *video mapper* takes motion cues from a pixel image as shown in figure 14.13. Three motion cues are determined: First it counts all visible pixels and determines the bounding rectangle defining the area of the picture that contains all non-zero pixels. The instant width and height of the bounding rectangle are computed and their peak-to-peak amplitude variations constitute the cues width-p and height-p. Second, the maximum velocity of gestures in the horizontal plane is calculated, followed by a calculation of the time between gestures in the horizontal plane—the third one.

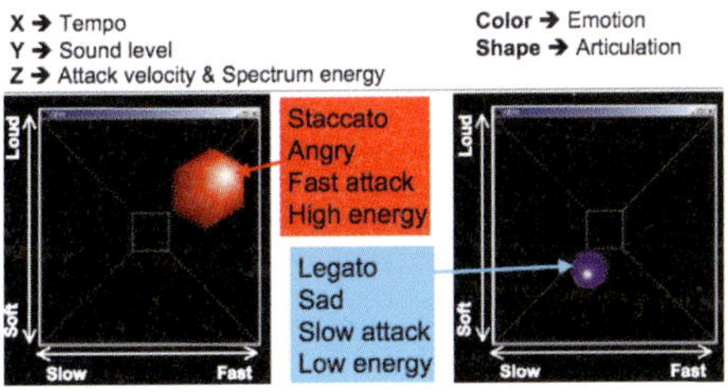

Fig. 14.12. Antonio Camurri's expressive mapper visualization coined *expressiball*.

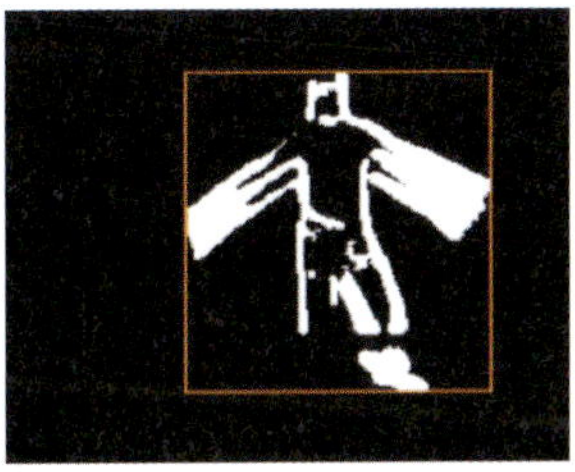

Fig. 14.13. The video mapper input takes essentially the rectangle around the visible pixels.

Players used the system in the 2004 group game Ghost in the Cave at the Tekniska museet Stockholm (figure 14.14). The game is played in two teams, each with a main player. Each team's task is to control a fish avatar in an underwater environment, moving it into three different caves. In the caves are ghosts expressing different emotions. The main players have to express the same emotion, causing their fish to change accordingly (figure 14.14).

Fig. 14.14. The Ghost in the Cave game at the Tekniska museet Stockholm.

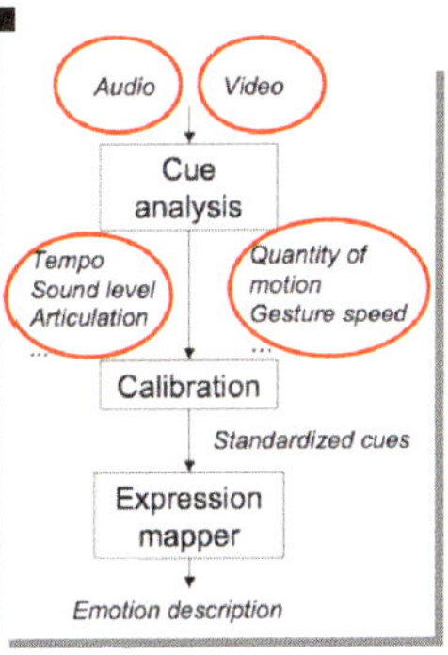

Fig. 14.15. The audio-video processing flow chart.

15

Analytical Expression

All words,
And no performance!
Philip Massinger (1583-1640)

This last chapter dealing with expressive performance focuses on analytical rationales of performance, which means that the shaping of performance is grounded in analytical facts, as opposed to gestural or emotional ones. This means that we have to analyze the given score's structure and then define performance by use of such facts. It is evident that this is the most explicit approach and also the easiest, since analysis of a score yields very precise and detailed information and, in contrast to gestures and emotions, has a scientific tradition, which offers good models to deal with when it comes to performance identity.

Nonetheless, musical analysis is not automatically translatable to performance, because traditional analysis is not formulated in commands for performance. For example, the fact that we have a chord whose Riemann function is a tonic in $C_{\#}minor$ does not automatically yield a quantity of loudness or a specific shape of the tempo curve on that chord. Therefore, analytical expression still needs a transformation process from traditional symbolic analysis to suitable performance commands.

We shall discuss three approaches to analytical expressive performance: Adorno, Friberg-Sundberg-Fridén, and our own approach. While Friberg-Sundberg-Frydén and Mazzola use computers and are forced to display completely explicit concept frameworks and rules, Adorno's approach is interesting in that it claims to be utterly precise, but in fact is not. It is not in the sense that a performance that follows Adorno's discourse is by no means well defined. We shall see to what extent this is so. This makes clear how difficult it is to step from analysis to performance when it comes to the details of performative structure. We could also have chosen a text by Schenker, but that other major

G. Mazzola, *Musical Performance*, Computational Music Science,
DOI 10.1007/978-3-642-11838-8_15, © Springer-Verlag Berlin Heidelberg 2011

theorist dealing with analytical performance, the picture would have been the same as with Adorno. This conclusion confirms Danuser's statement that at present (more precisely, this was 1992; by now, we have made some progress, as also testified in this book's title) we do not have a comprehensive performance theory [21, p.320].

15.1 Adorno's Analytical Performance Study of Webern's First Bagatelle

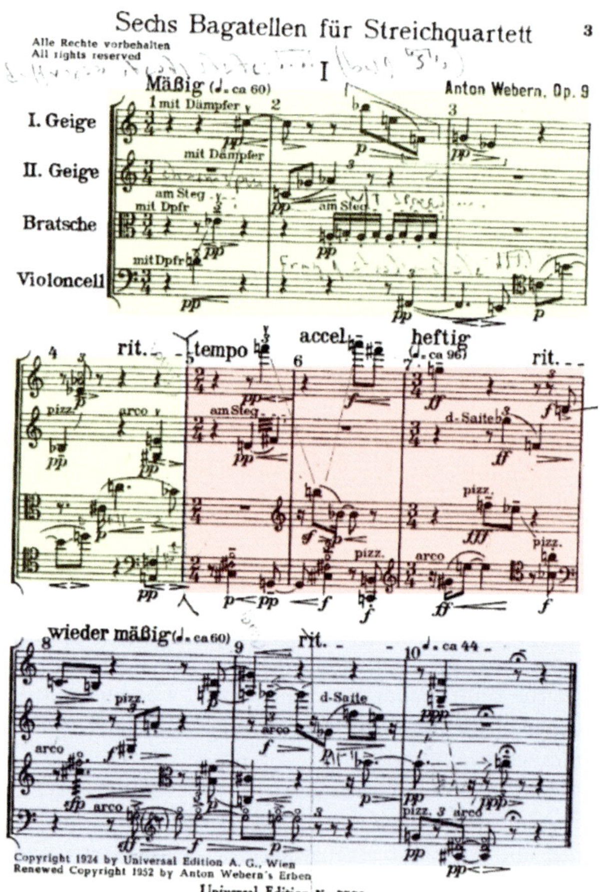

Fig. 15.1. Bagatelle no. 1, m.1-4: exposition; m.5-7: development; m.8-10: recapitulation.

Anton Webern's six Bagatelles op.9 were composed in 1913, so they pertain not the dodecaphonic composition technique but to a period, also called "free atonal," around 1909-1923. Dodecaphonism was only invented in 1921 by Arnold Schönberg (and discussed with his friends in 1923), the first dodecaphonic composition being probably Klavierstück op.23.5, a waltz from 1923.

We shall now discuss Adorno's analysis of Webern's first bagatelle (♫ 15), written in 1963 in Adorno's famous treatise "Der getreue Korrepetitor" (The Faithful Correpetiteur) [1]. This text deals with performance of works of the second Viennes school: Webern, *Lieder* op.3 and op.12, *Sechs Bagatellen für Streichquartett* op.9, *Vier Stücke für Geige u. Klavier* op.7; Schönberg, *Phantasie für Geige mit Klavierbegeitung* op.47; and Alban Berg, *Violinkonzert*.

Before we delve into Adorno's text, we should try to see what can be said about the first Bagatelle in terms of neutral analysis, i.e. of the analysis of what's objectively there, independently of heavily interpretational perspectives. Figure 15.1 shows the score. It has a tripartition, which is also observed by Adorno: m.1-4 = exposition, m.5-7 = development, and m.8-10 = recapitulation, although it is not evident that these titles taken from sonata theory should be applied with the given meaning in this microscopic context. There are, of course, a number of indexes for this partition, among them the rit./tempo and 3/4 to 2/4 time signature change on measure 5, as well as the second tempo change rit./wieder mäßig in measure 8. We accept them just as a general framing.

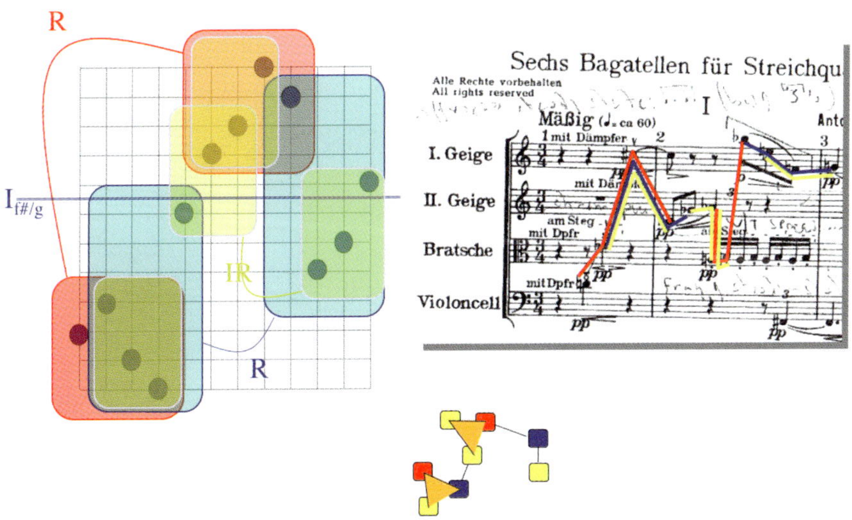

Fig. 15.2. The chromatic pitch distribution in the first two measures, together with the prominent inner symmetries of parts ("charts" of the covering) and the intersection configuration (middle, lower part), the so-called nerve of the covering.

The beginning of the piece shows a perfect 12-tone row (ante rem), see figure 15.2. We also recognize strong inner symmetries of that row. The graph in the middle lower part shows the "nerve" of this covering of the row by the eight charts, i.e. the intersection configuration: lines between intersecting charts, triangular surfaces between triples of intersecting charts. There are only three different isomorphism types of charts: the yellow with three elements, the red with four elements, and the blue with four elements. The beginning of the piece shows a perfect 12-tone row (ante rem) (figure 15.2).

In measure 4, the last measure of the "exposition," we see another highly symmetric construction (figure 15.3). The measure brings a sequence of four groups of notes that are all expressing the inversion symmetry $I_{f\#/g}$ in the pitch class set $\mathbb{Z}_{12}$. The concluding four-note group played as a chord by violoncello and violin II is also a nice symmetry between these two instruments: One plays the $I_{f\#/g}$ inversion of the other's notes. Interestingly, these notes all are different, so they build ten out of the twelve pitch classes. And the missing oncs, $f_\#$ and g, are the ones defining the inversion symmetry $I_{f\#/g}$. The symmetry-generating pair $f_\#, g$ reappears again in the last interval of violin I in the last measure 10. So the symmetry that concludes the exposition also reappears in the final interval, and this is also confirmed by the other final notes in measure 10: c and $c_\#$, the tritones to $f_\#$ and g, again confirming the same inversion. Recall that it is also the inversion involved in the retrograde inversion of the three-element chart in the initial twelve-tone configuration (figure 15.2).

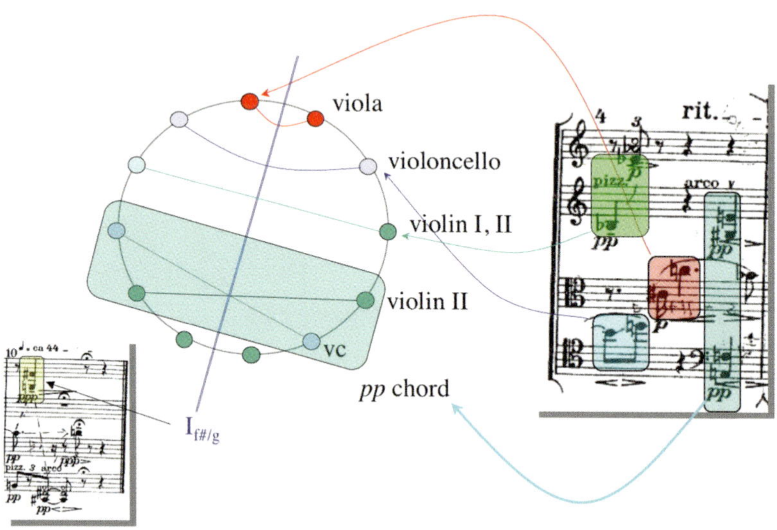

Fig. 15.3. The chromatic set and its symmetry-driven instanciationin measure 4.

So we expect that an analytical performance of this piece should take care of the above facts if it reconizes them. Let us now look at Adorno's text and let us always ask whether he connects the analytical discourse to the shaping of performance. We shall not discuss all the details here, but just a selection, which is significant for the style of Adorno's "precision."

The following text fragment from [1, p.284-287] is our translation. We include numbers for the paragraphs in order to have a better orientation.

> *(1) The relation of the first piece to the quartet setting of classical Viennes type is a double: on the one hand in its articulation in exposition, complication and coming back, then also in the part setting, the change of the main voice from instrument to instrument. The articulation following the movements is visible via tempo modifications. To the exposition correspond the first four measures, showing a connecting ritardando at the end.*
>
> *(2) The developmental part is separated from the exposition by a short general pause. This connection must be realized following the dynamic sonata-type compositional setting. One may connect to the preceding by the fact that between the benchmark notes of the former main voices, d in cello and c in viola, the chromatic intermediate note c-sharp is missing. This note is played by the syncopated flageolet entry of the cello at the beginning of the development, sounding two octaves higher than the fixed tone. Moreover, a moment of syncopation is inherited from the preceding phrases, which now, urged, traverses the measures of the development. The critical c-sharp must really be heard in the melodic context with the d and the c. But this c-sharp being a flageolet note is very different in its color from the other two critical tones. Therefore one should start earlier with playing towards this benchmark note by the cello and then by the viola. This is also suggested by the dynamical indications. However, after a minimal* crescendo *and* diminuendo, *the cello's d should be a nuance softer than the syncopated c of the viola. Consequently, the flageolet c-sharp of the cello must also be stressed a bit. This is sufficient in order to clarify the connection, supposing that the accompanying voices are really stepping back. The viola voice in the fourth measure must be understood as a model of syncopation in the development. With similarly spared strong rhythmical positions follow the entry of the cello, the viola, again the cello and the second violin. These entries combine to a melody. To the melody that is added from the syncopated phrases, on the climax, so-to-speak as a rhythmical resolution, the non-syncopated main voice of the first violin is added,* forte *and* crescendo. *The rhythmical difference between this phrase and the syncopated ones must be extremely concise.*
>
> *(3) The recapitulation phrase, three-measure long like the development, starts with the upbeat of the first violin, end of measure 7. It is denoted by* forte; *but one should not exaggerate the forte, it must be significantly weaker than the preceding fortissimo in the same measure. The* ritardando *however starts before that upbeat; the* pizzicato *chord*

of the cello is included, the b of the first violin would be understood as being again within the a tempo. *The middle phrase vibrates within the entire quasi-recapitulation. We start with an agitated tremolo, whose equivalent was heard in the beginning of the development: one should feel the similarity. Above all, the idea of syncopation, much as the flageolet sound, is fixed in the quasi-recapitulation; the tone repetitions of the cello correspond to the syncopated repetition of the c-sharp in the beginning of the development. That d of the cello must intervene so-to-speak without respect to the dynamics of the remaining phrase. The last measure of the development and the beginning of the quasi-recapitulation is the turning point of the piece. Of course the explosion of the cello must require just a moment, such that the attention can be driven upon the thematic main voices.*

(4) The second procedure, as conserved from the Viennese classicism, the exchanging of main voices from one instrument to the other, needs a special diligence in the performance. Webern's music is thoroughly shaped in each tone, every one is necessary in such a way that this change is not as automatic as it is essentially realized with Beethoven, where the musical language takes care a priori of our recognition of the theme. One has to pay attention to the completion of the melodic voices, to the right proportions of the partial motives which enable a single, however broken line. No holes are admitted in the melodic continuum; the new voice takes over, as they say, the old one.

(5) In the study of the third measure, the onset of the high cello voice must be realized immediately after the end of the e-flat of the first violin. The c-sharp of the viola in the fourth measure is played before the d of the cello has disappeared; this densification of the relation between the voices must be stressed.

(6) Moreover we have to pay attention to the inner life of the melody that is spread among the instruments. It is built in such a manner that the partial phrases are first dilated and then compressed. By means of such proportions the exposition is divided in an antecedent and a consequent; this articulation must be modeled. If viewing the exposition as a theme, it would be such that it successively unfolds from the departure. The first is a tone that grows out of the accompanying notes of cello and viola, the c-sharp of the first violin. A three-note phrase of the second violin does continue this one (measure 2): then the melody is prolonged in the first violin to five notes. Simultaneously with this expansion a small accompanying system built from sixteenth notes is formed in the viola, and finally appears a melodic counterpoint in the cello.

(7) The shaping strategy would be much too primitive within such differentiated melodic and harmonic relationships if it just consisted of a take-over of one main voice by another. The cello counterpoint f-sharp, which is added at the ending of the longest main phrase of the first violin, is already quite autonomous, containing its own melodic

life. Thereby the following cello main voice is prepared, which must be separated from the counterpoint by diligent phrasing.

(8) The dynamical indications, which give the cello counterpoint its life, testifies the intention of a transition to the foreground. The performance has to account for the crossing of voices, an most important technique of concatenation, where the main voice, here in the first violin, is dropped at the end, whereas the secondary voice of the cello becomes visible by the crescendo, and then immediately is taken back. In this spirit, this type of performance must be transported to crossing locations, of course in a modified manner, according to the specific situation. Once these moments have been elaborated, they appear to be exaggerated; on the next level of the study, they have to be dampened. In an analogous way, the last two measures of the piece are presented where the main melody of the second violin is diminshed, while the viola is added slightly more stressed, and then, including the melodic cello flageolet c-sharp, fades away. The critical notes are similar to those which lead from the exposition to the development; on this is based the effect of the recapitulation, and those tones should be stressed, still maintaining the tenderness, in such a way that the similarity is apparent.

(9) With such a subtle compositional approach as it is realized in the bagatelles, minimal compositional means are sufficient, such as the identity of three notes, or the simple crossing of voices on two corresponding positions, without any motivic relation, in order to create correspondences. Let me point out one more such correspondence to the performers. There are two homophonous chords in this piece, added from double stops on two instruments: at the end of the exposition (measure 4) and in the recapitulation (measure 8). Not only have these chords to be played very clearly in order to stress their correspondence, following Webern's requirement, one also has to play a beautiful sound. Such exceptional situations like these quasi-homophonous chord sounds must be heard as such; they must so-to-speak be savored. We have to feel all the sensuality that stands behind Webern's music, which was living in the Passacaglia, and which, as a negated one, is still secretely present. They report that Webern had asked a famous conductor to play a beautiful sound in an orchestral piece. This one answered, "I am not a rubato conductor." I believe that one should not be more papal than the pope, not more ascetic than Webern. The correct performance of his pieces needs a sensual freedom despite the rigor, without destroying the context. Webern went quite far in this direction; but such freedom is legitimate only as a result of extremal precision.

The first performance-related statement comes in paragraph (2), where Adorno suggests that one should connect to the missing $c_{\#}$ between the benchmark notes c, d. However, that $c_{\#}$ is played by the viola in measure 4, so why is it missing? And then, how do we have to perform that $c_{\#}$, which now is played in measure 5 by the cello? Adorno recommends by some arcane argument of flageolet sound color of the $c_{\#}$ that one should play the d and then the c "earlier towards this benchmark note." Earlier, yes, but then Adorno refers to the dynamical signs as a suggestion to play earlier. This makes no sense as a performance command. The precision is more in Adorno's imagination than in the concrete performance.

In paragraph (3), Adorno asks the first violin to play its b in a *forte* that is "significantly weaker than the preceding *fortissimo* in the same measure." A *forte* is always significantly weaker than a *fortissimo*. So what is the performance command here? And if so, why must it be played "significantly" weaker? No analytical reason is given here.

In paragraph (4), Adorno stresses the shaping of each tone as opposed to the automatisms with Beethoven, for example, where the "musical language takes care a priori of our recognition of the theme." Nice thought, but then, what should we do? "...Pay attention to the completion of the melodic voices, to the right proportions of the partial motives which enable a single, however broken line." Still nice, but everything except precise. The thought does not specify the performance. If a line is single, but broken, how should I play *this* insight?

In paragraph (5) we excepionally have a more or less concrete performance command: Play that $c_{\#}$ before the d ends. And the analytical reason is "densification of the relation between the voices."

In paragraph (6), again, Adorno exposes some nice phantasies without any concrete performance commands. And the phantasies are everything but analytically valid. All those insights, which we have described above, are absent; instead, he invokes the "inner life of the melody." Nice, but poetic hot air.

In paragraph (7), the performance command reads as "...must be separated from the counterpoint by diligent phrasing." No precision whatsoever.

In paragraph (8), the final recommendation—"those tones should be stressed, still maintaining the tenderness, in such a way that the similarity is apparent"—is beautifully fuzzy. Tenderness, a feeling, perhaps a haptic category, but how would one perform this, and then in what way that the similarity is apparent?

In paragraph (9), the negation of the Passacaglia, however still secretly present, what can that help in performing concretely? It sounds cynical to claim that Webern's freedom is obtained from extremal precision, whereas this text is only claiming precision and never gets concrete with respect to the objectively present structural richness described above.

How should one play that inversion, those repeated/transformed motives in the initial row, those symmetric tone groups in measure 4, and their dominating inversion $I_{f_{\#}/g}$ made explicit in the pitch combinations of the concluding measure? We have the impression that Adorno would like to be precise, but he stops this enterprise on an aesthetic level of fuzzy poetical invocation instead of precise shaping of performance. His performance in analysis is insufficient for analytical performance.

15.2 Anders Friberg, Johan Sundberg, and Lars Frydén: Director Musices Performance Software

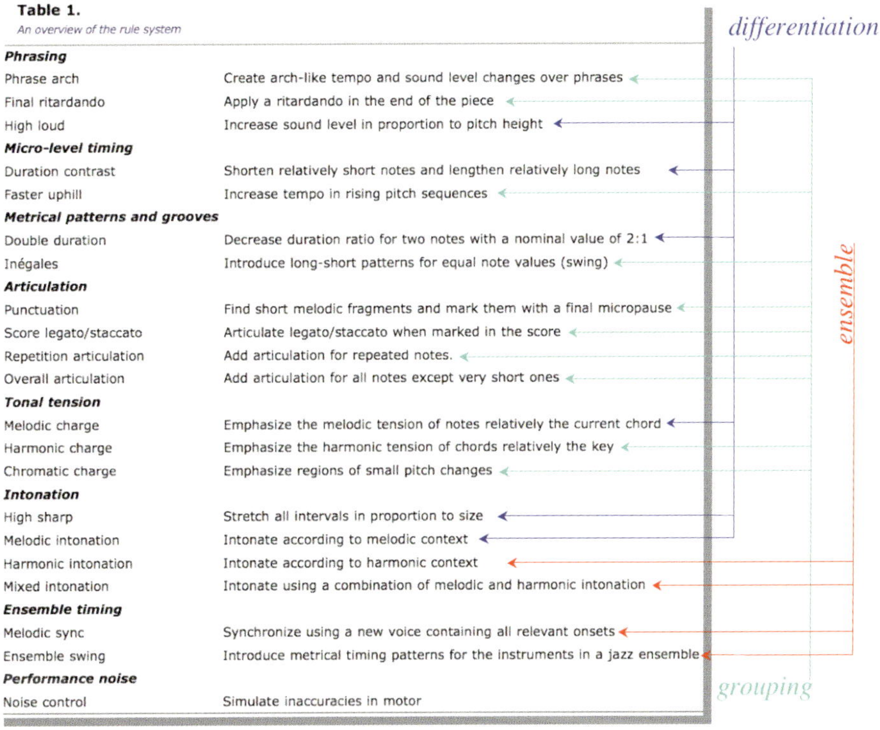

Table 1.
An overview of the rule system

Phrasing	
Phrase arch	Create arch-like tempo and sound level changes over phrases
Final ritardando	Apply a ritardando in the end of the piece
High loud	Increase sound level in proportion to pitch height
Micro-level timing	
Duration contrast	Shorten relatively short notes and lengthen relatively long notes
Faster uphill	Increase tempo in rising pitch sequences
Metrical patterns and grooves	
Double duration	Decrease duration ratio for two notes with a nominal value of 2:1
Inégales	Introduce long-short patterns for equal note values (swing)
Articulation	
Punctuation	Find short melodic fragments and mark them with a final micropause
Score legato/staccato	Articulate legato/staccato when marked in the score
Repetition articulation	Add articulation for repeated notes.
Overall articulation	Add articulation for all notes except very short ones
Tonal tension	
Melodic charge	Emphasize the melodic tension of notes relatively the current chord
Harmonic charge	Emphasize the harmonic tension of chords relatively the key
Chromatic charge	Emphasize regions of small pitch changes
Intonation	
High sharp	Stretch all intervals in proportion to size
Melodic intonation	Intonate according to melodic context
Harmonic intonation	Intonate according to harmonic context
Mixed intonation	Intonate using a combination of melodic and harmonic intonation
Ensemble timing	
Melodic sync	Synchronize using a new voice containing all relevant onsets
Ensemble swing	Introduce metrical timing patterns for the instruments in a jazz ensemble
Performance noise	
Noise control	Simulate inaccuracies in motor

Fig. 15.4. The rule system in the KTH system [37].

Friberg's, Sundberg's, and Frydén's approach was conducted at Stockholm's Kungliga Tekniska Hgskolan (KTH). In 1982 and 1983, Sundberg et al. presented a set of eleven rules constituting the start of the rule system [131]. The system was written in Lisp and called *Rulle*, later then *Director Musices*. The system has been continuously developed and can be downloaded from [38].

In 1991, Friberg published the paper in the *Computer Music Journal* [35] with all the classical rules, entitled *Generative Rules for Music Performance: A Formal Description of a Rule System*. On his homepage [39], Friberg has an number of sound examples for his rules.

The authors describe the rule system as follows [37]: "The aim of the rule system is to find general principles of music performance." The only system parameter for a rule is a number k: $k = 1$ for default, $k > 0$ for positive action of the rule, $k < 0$ for negative action ("opposite" of the rule). Rules can be combined to a palette of rules for all kinds of macro actions, like stylistic types. The rules act if a specific context is discovered, and then are executed. All rules act on given physical parameters. The system of rules is shown in figure 15.4. They are grouped in three categories: differentiation, ensemble, and grouping [37].

Let us look at some of them.

- **Rule DPC 1B. High Loud**
  The sound level of tones is raised by $3k$ dB/octave:

  $$\Delta L = \frac{N - 60}{4} \cdot k \text{ [dB]}$$

  N is the semitone number with $N = 60$ for note C_4.
- **Rule GMI 1C. Faster uphill**
  The duration of a tone is shortened by $2 \cdot k$ msec if the preceding tone is lower and following tone is higher. This shortening is also applied on the first tone in an unbroken series of ascending intervals.
- **Rule GMA 3. Final Ritard**
  This rule is optional. Let T_n be the running time from the beginning of the ritard to the current one, index n, and let T_{tot} be the total length of the ritard. The time T_n is taken at the middle of each tone, i.e., at $DR_n/2$. Then, the change in duration of the note in the ritard will be

  $$\Delta DR_n = \frac{100}{\sqrt{1.1 - \frac{T_n}{T_{tot}}}} \text{ [\%]}$$

  (Maximum 330 percent at the last tone)
  Let us look at an example of this rule, as published online [40]: J. S. Bach, Invention for two voices, F major, BWV 779. Here two other rules were also applied: the higher, the louder, increasing loudness with increasing pitch, and punctuation, inserting micropauses after melodic gestures. The figures 15.5, 15.6, 15.7, and 15.8 show four versions with variable system constant k. See ♫ 16 with the sound examples for these cases.

The method used to find these rules is called *analysis by synthesis*. This means that a rule is first defined provisionally, and next applied to a number of musical situations. The effect of the rule is then critically reviewed by a performance expert, professional violinist Lars Frydén in the case of the KTH system. He proposes changes of the rule, and so on. So it is a trial and error system. But we have to recognize the innovative step taken in this endeavor. In view of the preceding approaches to analytical performance, the KTH group

Fig. 15.5. $k = 0$, no ritard.

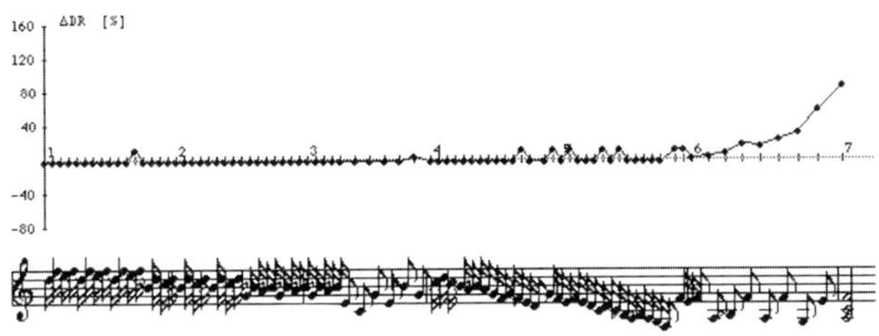

Fig. 15.6. $k = 1.3$, medium ritard.

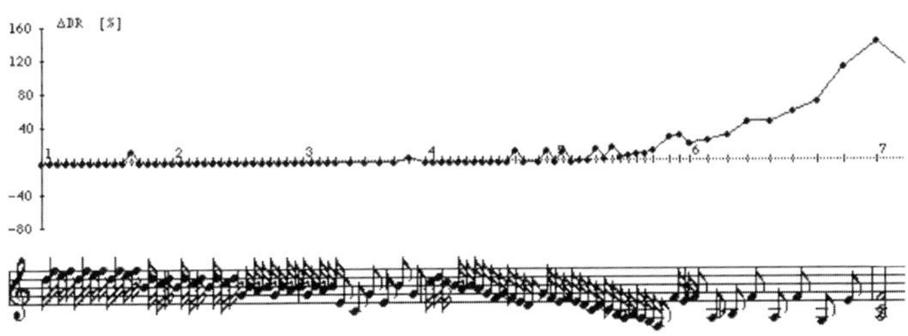

Fig. 15.7. $k = 2.1$, exaggerated ritard.

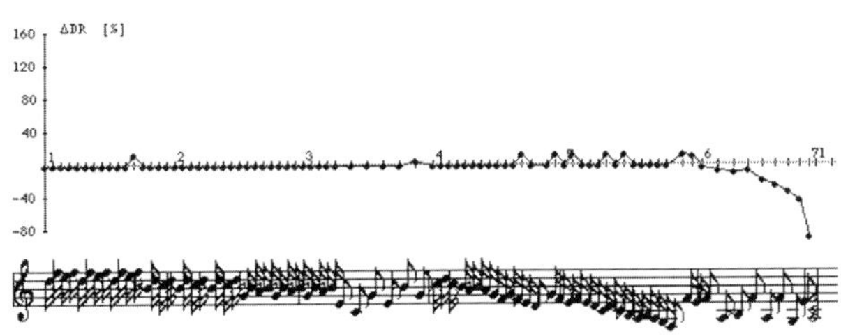

Fig. 15.8. $k = -1.3$, inverted ritard.

was literally in the empty space with the rules to be established. No concrete rule was made available by those music or professional performance theorists. Adorno's 'romantic' precision was of no help either, as we saw above. Moreover, the claim of the KTH group was also extremely demanding: to define *general* rules, which means that they apply to any given score within a wide range of styles, e.g. composed within traditional Western harmony (to apply the rules concerning melodic and harmonic charge).

The question arises here whether a more specific and refined analysis would be necessary to yield adequate performance commands. The gap is evident: No refined analysis of rhythmical, melodic, contrapuntal, or harmonic structures is used in these rules. In particular, the local variation of tempo, agogics, and the note-related dynamics or articulation are not considered as functions of specific local analysis of the score's structure. But this is what an expert performer would have to consider: How do I play the particular harmonic situation in that determined chord? Do I have to play a slight ritardando? Do I have to stress the loudness of a harmonically important note of that chord? Would I have to play the chord slightly staccato to detatch it from subsequent harmonies?

15.3 Guerino Mazzola and Oliver Zahorka: The Rubato Theory of Analytical Performance

Fig. 15.9. Oliver Zahorka.

Knowing about the KTH rule system and also with regard to the general structure theory of performance, which we have described in part II of this book and which was first sketched in [79], the author and his highly talented coworker Oliver Zahorka collaborated on a project funded by the Swiss National Science Foundation in 1992-1996 on performance theory and its computerized implementation, see also [87] for a first published report on the project. We come back to the implemented software and its history to present in part IV.

The general idea was to design a meta-theory and implement a software in the sense that no fixed rules would be given but a modular scheme of specialized components for musical analysis and performance. So the expressivity was strictly limited to the analytical approach, but neither analysis nor performance was 'hard-coded.' This enterprise however had to be shaped in a generic process scheme. This process scheme then gave rise to the RUBATO® software. Basically, we had three logical components of this system:

1. The data format used for internal processing of musical compositions.
2. The analytical components taking the musical data from 1. and transforming them into some analytical output data.
3. The performance component with its variety of shaping procedures, producing an audible output using the analytical output from 2.

Component 1 turns out to be a hard problem. We want to be able to process any reasonable music data and to represent them in our data format.

The difficulty arises not from the existing formats for music data, but from the important fact that musicians and composers could at any time come up with new concept architectures for their productions. We therefore cannot rely on a given database format, which is always hard-coded. We have solved the problem by the invention of the dynamic concept architecture of *forms* and *denotators*. Forms are kinds of general, recursively defined spaces, while denotators are points in those spaces. The denotator framework is extensible—one can always add new concepts if required. The flexibility comes from the general approach to concepts by modern mathematics, specifically from topos theory, and from some paradigms of object-oriented programming. We do not discuss this topic here but want to stress that denotators are still the basic data format in Rubato's present architecture, see [97, Chapter 5] for details.

Component 2 takes the musical composition that originally was fed into the system in the now transformed format of a denotator of notes, pauses, and other signs. This component must perform an analytical task according to any given theory. We are not making decisions about the chosen approach. The only condition is that the output be in a format that makes sense to performance shaping. One might think of rhythmical, melodic, harmonic, contrapuntal analyses. Or analyses of the large form, including phrasing, beginning, and ending analyses, as needed for the final ritard operator in the KTH system, for example. The selection is not important. The theoretical background may be Riemannian, Schenkerian, it does not matter. What is important, however, is the output format. The philosophy here is that whatever comes out from the analytical processing should be usable directly for performance. Performance is a shaping of that deformation as described in all details in part II. Therefore, such deformation being a geometric, parameter-defined operation, the performance commands need to act numerically in one way or another. More practically speaking, the musician has to execute numerical actions when performing. The keys must be hit more or less, the violin bow has to be moved more or less fast, the pressure on the violin's strings has to be more or less strong.

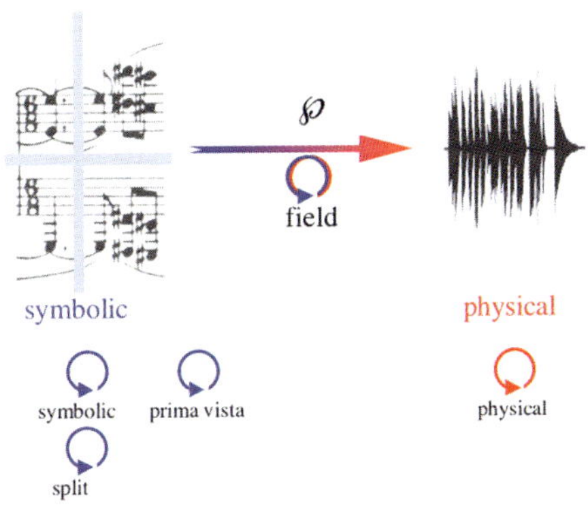

Fig. 15.10. Shaping operators can be symbolic and act on the symbolic kernel, as for example the split operator, then certain features of the primavista operator, or the symbolic operator, which applies weights to parameters of the symbolic kernel. The field operators are those, such as the tempo operator or the scalar operator, that act on the given transformation $\wp$. The physical operator applies a weight to the given output on a determined set of output parameters.

A second argument for such a choice is that we would also like to use analytical output by any operator that would shape performance. *All operators should be able to use any given analysis.* We did not want the performance component to be forced to look at specific analytical output data before really stepping to the performative core business. A musician does not want to worry about the analysis' text, but just use it to do the performance shaping—to move the hands, to blow stronger, to move the bow faster. Therefore, it was decided to apply this insight as a general philosophy for the processing of analytical output in the performance component. This has been achieved by the requirement that the output of any analytical processing be what we coined an *analytical weight*. Such a weight is a function of the composition's notes, and each note takes a numerical value.

Component 3 deals with the actual creation of a performance from the analytical weights received from component 2. This means that this component has to provide us with any type of *shaping operator* that defines a performance transformation. The first delicate thing is that such a transformation has three parts: the symbolic input, the transformation map $\wp$ itself, and the sounding 'physical' output. So an operator might act upon each one of these three perspectives. Accordingly, such operators are called *symbolic*, *field*, and *physical* (figure 15.10). For example, a symbolic operator would be the splitting of right and left hand in the construction of a performance. Shaping tempo would be a field operator, and changing the given sound duration by some percentage for articulation would be a physical operator. This distinction is quite significant when compared to the KTH rules, which are all physical, i.e. they always act on the sounding output.

A second even more delicate aspect of performance operators comes into play here, an aspect which will lead us very far into the intrinsic logic of performance: *any operator has to act upon a given performance.* This one might be the primary mechanical sight-reading performance, or a more sophisticated one. In fact, one never knows when in the trajectory of performance shaping a determined operator is to be applied. This principle has deep consequences in the construction of performance operators, because one has to define them in order to be capable to intervene at any stage of the shaping process. We come back to this problem in the discussion of the stemma theory of performance in section 21.

16

Analytical Weights

Weight and measure save a man's toil.

Analytical weights did not come up from empty space. In fact, our idea was taken from Hugo Riemann's definition of metrical weights: Meter relates to weights. So it was decided to generate an output in the form of numerical weights for any analytical engine. Here is the precise definition of a weight:

Definition 1 *An analytical weight is a continuous function*

$$w : PARA \to \mathbb{R}$$

defined on a space PARA of parameters such as $\mathbb{R}^E$, $\mathbb{R}^H$, $\mathbb{R}^{EHL}$, *etc., with non-negative real values.*

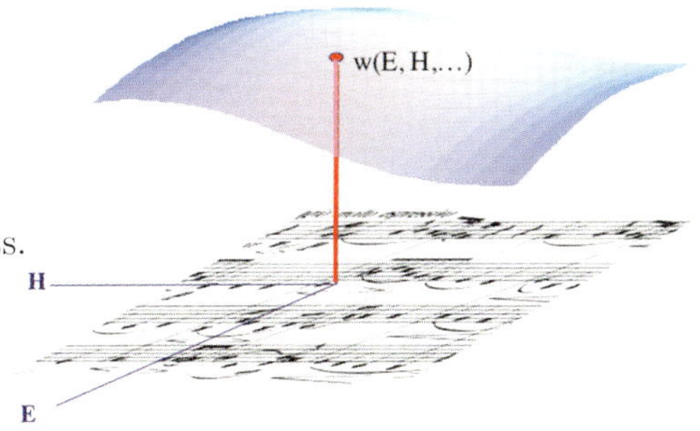

Fig. 16.1. An analytical weight on the space $\mathbb{R}^{EH}$.

Such weights are calculated upon music analyses and correspond to associated semantics. For example, a metrical weight $w : \mathbb{R}^E \to \mathbb{R}$ might associate metrically important onsets with higher weights. We observe that weights are also defined where no notes are present. This is not a restriction, since any discrete functions defined only on notes, say, can be extended in a continuous way to the entire space. There are also deeper reasons for this setup. Since performance fields are defined on entire frames (see section 10.1), their shaping must be defined for any argument of those frames, also where there

G. Mazzola, *Musical Performance*, Computational Music Science,
DOI 10.1007/978-3-642-11838-8_16, © Springer-Verlag Berlin Heidelberg 2011

are no notes. Therefore, it is wise to use weights that are defined all over the place. There is also a more computer-driven argument for such an extension. If we are to apply a shaping to a given note, it might be that for certain computer-generated imprecisions, the position of the note cannot be identified with the position of a discrete weight, when applied to that note. Therefore, it is prudent to extend the discrete weight continuously to the neighborhood of each note point.

We have implemented the construction of continuous weight functions from discrete weights by use of cubic splines. Cubic splines are uniquely determined cubic polynomial functions $P(x) = a_3X^3 + a_2X^2 + a_1X^1 + a_0$, which connect two values f_0, f_1 at two arguments x_0, x_1, respectively, with the given slopes s'_0, s'_1. This means that $P(x_0) = f_0, P(x_1) = f_1, P'(x_0) = s_0, P(x_1) = s_1$. This construction can be extended by recursive procedures to functions on higher-dimensional spaces [84, section 32.3.2.1]. Our slopes are always set to zero, so that the local variation of the continuous extension is minimal, if the argument is slightly different from the required data.

In the following section, we present a bunch of analytical tools, which were implemented in the Rubato software to give prototypical examples of analytical procedures following the above weight philosophy. Although none of these was thought to be a particularly creative contribution to musical analysis, it turned out that they all quite ironically entailed successful scholarly careers of those specialists[1] who delved into these analytical topics without deeper connections to performance theory as such.

16.1 Metrical Weights

The metrical analysis that we developed in this context can be understood from its central concept: the local meter (figure 16.2). This is akin to the one proposed by Jackendoff and Lerdahl [53], but differs in essential points: A local meter is a finite sequence $M = (E_0, E_1, \ldots E_l)$ of regularly distributed symbolic onsets E_i with constant interval $d = E_i - E_{i-1}, i = 1, \ldots l$, the number $l = l(M)$ is called *the local meter's length.* Local meters are however always built from onsets that appear as attributes of objects, such as notes, pauses, etc. in a score. Onsets that are not related to concrete objects are not admitted, in contrast to the approach in [53], and also in accordance with Riemann's understanding of metrical structure being supported by existing events.

A maximal local meter in a score is a local meter, which cannot be embedded in a properly larger local meter. Figure 16.2 gives an example of a maximal and of a non-maximal local meter. In the metrical analysis of a piece, we then project all notes to their onsets and look at the covering of those onsets by maximal local meters. Figure 16.3 shows a simple music piece X and its covering $Max(X) = \{a, b, c, d, e\}$ by five maximal local meters.

The notes are not all in the same position with respect to this covering. Some are contained in many maximal local meters, others in just one of them.

[1] This is Chantal Buteau for melodic analysis and Anja Volk-Fleischer for rhythmical analysis. And to a lesser degree Thomas Noll for harmonic analysis.

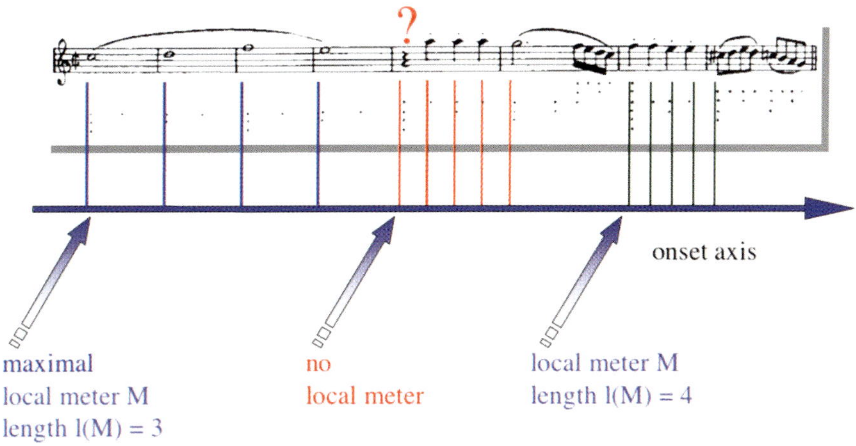

Fig. 16.2. Local meters: to the left a maximal one, in the middle a counterexample, to the right a non-maximal local meter. The metrical analysis is akin to the one proposed by Jackendoff and Lerdahl [53], from where we have taken the present score excerpt, the beginning of Mozart's *Jupiter Symphony*.

Some are contained in longer local meters, some in shorter. There are two views on this situation: a topological and a numerical. The topological one views notes of the composition X as being more or less dominant over others as a function of the maximal local meters which contain them. This is formally represented by the so-called *nerve* $\mathcal{N}(X)$ of the covering $Max(X)$. Figure 16.4 shows the situation.

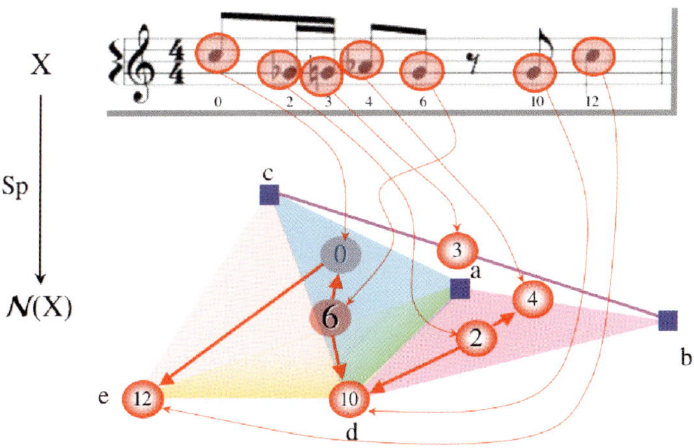

Fig. 16.4. The nerve of a composition X.

We have a map $Sp : X \to \mathcal{N}(X)$ that associates with each note $x \in X$ the set $Sp(x)$ of all maximal local meters containing x, this is called the *simplex of* x. We then see that certain notes are metrically dominant in the sense that they have larger simplexes than other notes. Musically speaking, this means that these notes participate in more local meters then others, so their metrical relevance is dominant. We see in our example that note 6 has a tetrahedron simplex—it is contained in maximal local meters a, c, d, e—whereas note 12 is only in the simplex that has a single maximal local meter e. So note 6 dominates note 12: That maximal local meter defines one vertex of the tetrahedron. Note 3 has a simplex built from two maximal local meters b, c, and we draw a line to visualize their common note 3. Note 2 has a triangle simplex: It is spanned by three vertexes, a, b and d.

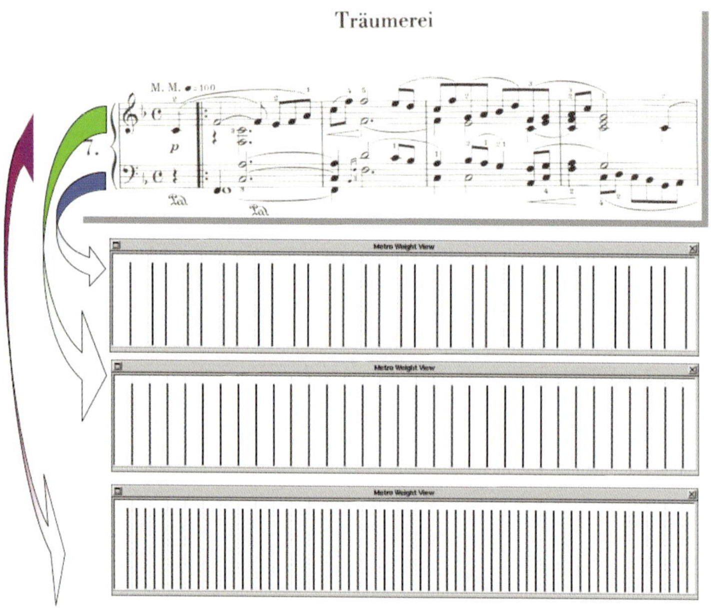

Fig. 16.5. The metrical analysis of Schumann's *Träumerei*, op.15/7, by Rubato's MetroRubette reveals a left hand weight of 3 +5 against a right hand of 4 beats, while the combined metric weight is an eight note offbeat metric.

The topological perspective is interesting, but far from what we expect: namely the weights associated with a given analysis. We would like to call the nerve the *global metric* of the piece, while the global rhythmic would be the following. We look at a given onset x. Then we look at all maximal meters containing that onset, and then we try to get a weight from that information. This information consists of two things: the maximal local meters containing that x, and then for each such object, a numerical value measuring this local meter's significance for that onset. We propose this formula for the metrical analysis we have implemented:

$$w(x) = \sum_{x \in M, m \leq l(M)} l(M)^p$$

where m is a minimal length of maximal local meters to be considered in this calculation, and where p (metrical profile) is a power that determines the relevance of lengths of local meters. The minimal admitted length m means that maximal local meters shorter than m are omitted. We only look at maximal local meters with sufficiently large length. We would call this function w the *global rhythmic* of X, the rhythmic being a function of the global metric structure but having a numerical expression, and this one being a weight function.

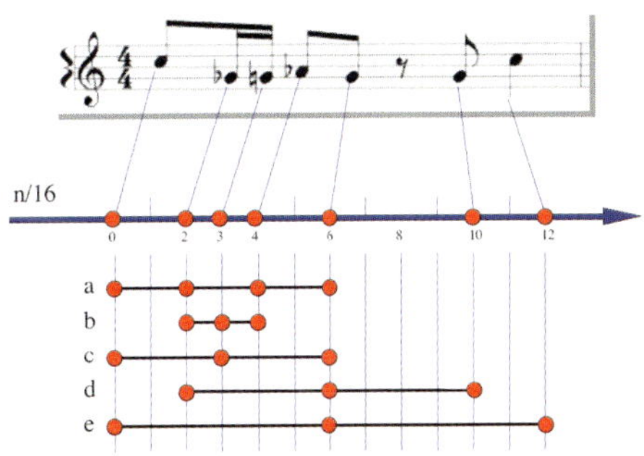

Fig. 16.3. The maximal local meters covering the piece X.

While this is a fairly satisfactory solution of the original problem, we are still left with some problems. In fact, in music scores, there are many different signs that relate to time: notes, pauses, notes from different instruments, bar lines, etc. How can we manage this? The approach is fairly simple. We take a number of such types of objects, like notes, pauses, etc. Let them be the types $t_1, t_2, \ldots t_k$. Then we have a weight function $w_i(x), i = 1, \ldots k$ for each of them according to the preceding theory. Further, we decide to give each of these weights a weight, i.e. a number $\nu_i \in \mathbb{R}$ measuring the strength of the weight w_i. Then we can define a combined weight by the function

$$w(x) = \sum_{i=1,\ldots k} \nu_i w_i(x)$$

Let us look at an example to illustrate the general technique and its usage. In figure 16.5, the right hand shows a regular 4-beat weight, when we go to the longest possible minimal length m where there are maximal meters. In contrast, the left hand shows a 3+5 structure; this means a two-bar regularity, a marked opposition to the regular right hand. This creates a strong tension in performance; perhaps this is felt by many pianists performing the *Träumeri*?! The weighted sum of both of these weights (with $\nu_1 = \nu_2 = 0.5$) shows a half-measure offbeat metrical weight. The left hand sound with its 3+5 structure can be heard from a computer-generated performance in example ♫ 17.

For a more detailed study of metrical weights, we refer to Anja Volk-Fleischer's work [33].

16.2 Melodic Weights

A second analysis that we have implemented was inspired by Rudolf Reti's work on thematic processes in music [113]. It turned out that here, much more than with the metrical/rhythmical analysis, there was no valid theory. The very concept of a melody or motif is not defined, and no precise theory about the body of motivic structures within a given composition is available. We do not discuss this dramatically underdeveloped theory here, but see [9] for a detailed account. Despite this deplorable state of the art of motivic analysis, we have initiated an analytical theory of motives in order to be able to implement such thoughts and to use them in the framework of the Rubato software.

To begin with, we suppose that the score is given as a set of events with onset and pitch and possibly some other coordinates. So they live in the space $\mathbb{R}^{EH\ldots}$. Then

Definition 2 *A motif M in $\mathbb{R}^{EH\ldots}$ is a finite set $M = \{n_1, n_2, \ldots n_k\} \subset \mathbb{R}^{EH\ldots}$ of k different notes having all different onsets, i.e. $E_{n_i} \neq E_{n_j}$ if $i \neq j$.*

With this definition, one may define different paradigms of similarity among motives, depending upon the information extracted from a motif's structure. For example, one may only look at the structure of increasing, equal, or decreasing successive notes. We omit the technical details here and refer to [84, Chapter 22]. Whatever is the similarity paradigm, we may then define precisely what it means when a motif is similar to, i.e. in a neighborhood of, another motif with the *same cardinality*. This is a concept of distance, so we may say that the distance $d(M_1, M_2)$ of motif M_1 to motif M_2 is less than 0.125.

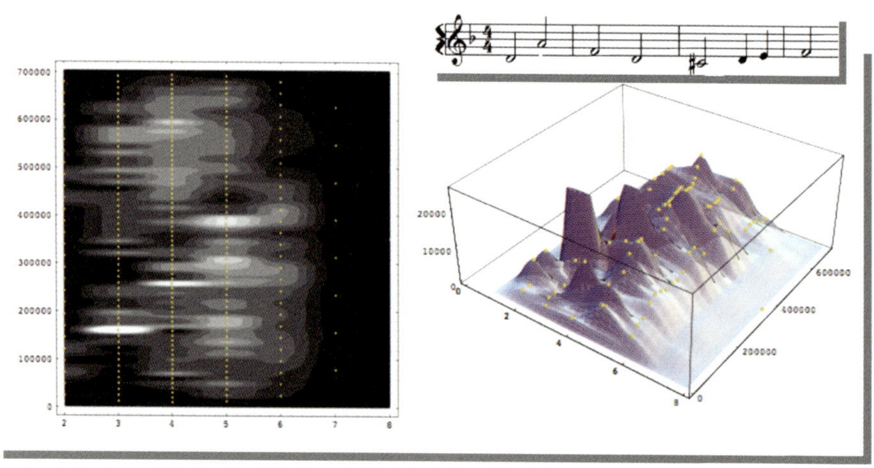

Fig. 16.6. The weights of motives in the main theme of Bach's *Kunst der Fuge*. The motives are grouped by their cardinality, and the weight to the left is encoded by brightness.

With these prerequisites, given a positive real number ϵ, we may define the melodic weight of motives (the definition is somewhat simplified here but gives the idea, see [84, Chapter 22] for a detailed account). Take a motif M in our composition and look at all motives N in the given composition such that there is a submotif $N^* \subset N$, with the same cardinality as M, such that $d(M, N^*) < \epsilon$. Call their number the *ϵ-presence* $pr_\epsilon(M)$ of M. Similarly, consider all motives L in the composition that are in the ϵ-neighborhood of a submotif of M, and call their number the *ϵ-content* $ct_\epsilon(M)$ of M. Then the *ϵ-weight* $w_\epsilon(M)$ of M is the product

$$w_\epsilon(M) = pr_\epsilon(M) \times ct_\epsilon(M).$$

So the weight of a motif 'counts' all motives that contain some motif similar to M or being similar to a submotif of M. This accounts for the motif's relation to other motives in the composition. See figure 16.6 for an example.

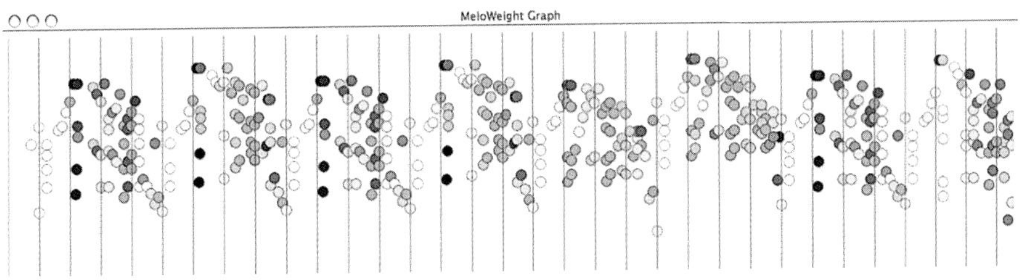

Fig. 16.7. The weights of notes in Schumann's *Träumerei* in Rubato's MeloRubette. The vertical lines are the barlines.

Given these numbers, we can define the *melodic ϵ-weight of a note x* of our composition to be the number

$$w_\epsilon(x) = \sum_{M, x \in M} w_\epsilon(M)$$

Its music-theoretical meaning is the account of all motives' weights, where x is a member. See figure 16.7 for an example, where the gray value of disks encodes the weight of the notes that are represented by these disks.

16.3 Harmonic Weights

The harmonic analysis that we have implemented is quite involved Riemann theory. Riemann harmony is designed to attribute to chords three types of harmonic values: tonic, dominant, or subdominant. Such a value is always related to the tonality where the given chord is situated. This valuation of chords generates a syntax of harmonic values, which reflects the harmonic semantics of tonal music. Harmony then makes statements about the harmonic meaning of given sequences of chords. Despite this fundamental role of Riemann harmony, Rubato's HarmoRubette for harmony is the rist to make Riemannian function theory fully explicit. The reason is that Riemann's harmony has never been completed because complex chords have never been given harmonic values by a reliable system of rules, but see [84, Chapter 25] for details. Our idea is this: We start with the sequence $(Ch_1, Ch_2, \ldots Ch_n)$ of all chords in a given piece X. We first calculate the Riemann function values for each chord. This means that for every Riemannian value $riem = T, D, S, t, d, s$ of major tonic T, dominant D, subdominant S, and minor tonic t, dominant d, subdominant s, and every tonic $ton = C, C_\#, D, D_\#, E, F, F_\#, G, G_\#, A, A_\#, B$, we calculate a fuzzy value $val_{ton,riem}(Ch_i) \in \mathbb{R}$ of chord Ch_i. This defines the *Riemann matrix* $val_{.,.}(Ch_i)$ *of* Ch_i. The fact that we do this in a fuzzy way means that we do not oversimplify the situation: It might happen that a chord is 'more or less' dominant in D major; this is the meaning of fuzzy values here.

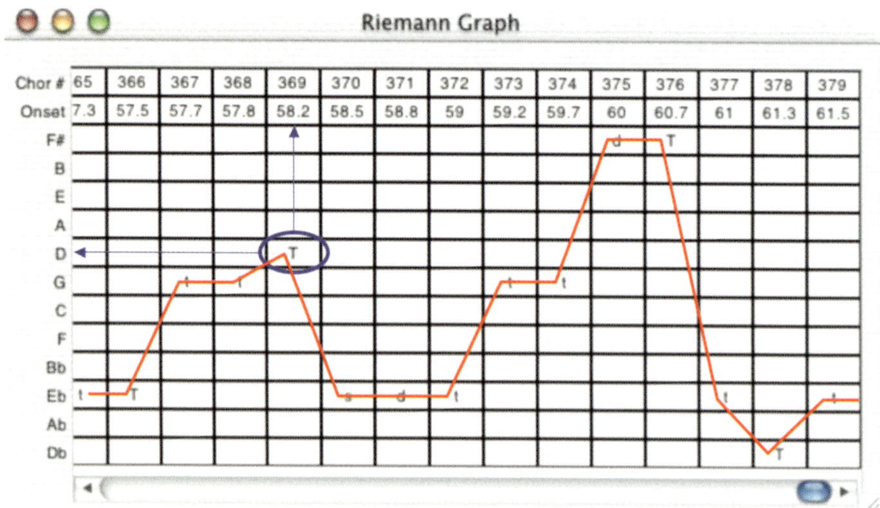

Fig. 16.8. The Riemann graph of a composition in Rubato's HarmoRubette. The sequence of chords is given Riemann values as a function of least transition weights.

Next, preferences allow us to set the transition weights for any pair of successive chords and Riemann parameters $(ton, riem, Ch_i), (ton', riem', Ch_{i+1})$, using also the Riemann matrix values $val_{ton,riem}(Ch_i), val_{ton',riem'}(Ch_{i+1})$. Harmonically difficult transitions will get larger weights than easier transitions. With this information, one then looks at all paths of Riemann parameters of chords

$$(ton_1, riem_1, Ch_1), (ton_2, riem_2, Ch_2), \ldots (ton_n, riem_n, Ch_n)$$

and calculates the weight of such a path as a function of the pairwise transition weights. The lightest path is then chosen as being a solution of the Riemann function attribution for all chords. Figure 16.8 shows an example of a harmonic path.

The calculation of harmonic weights of single notes is now easy (although very intense in terms of computer calculation work). We select a note x, living within a chord Ch_{i_0}. Then we calculate the weights for the chord $Ch_{i_0} - \{x\}$ and look at the ratio of the full path as compared with the weight of the path with the chord after omitting x. The weight of x increases if the ratio is large, and so we get weights of single notes. The technical details are described in [84, section 41.3], we omit them here.

Whereas the rhythmical weight is essentially a function on the onset space $\mathbb{R}^E$, both the melodic and the harmonic weights are functions on $\mathbb{R}^{EH}$.

16.4 Primavista Weights

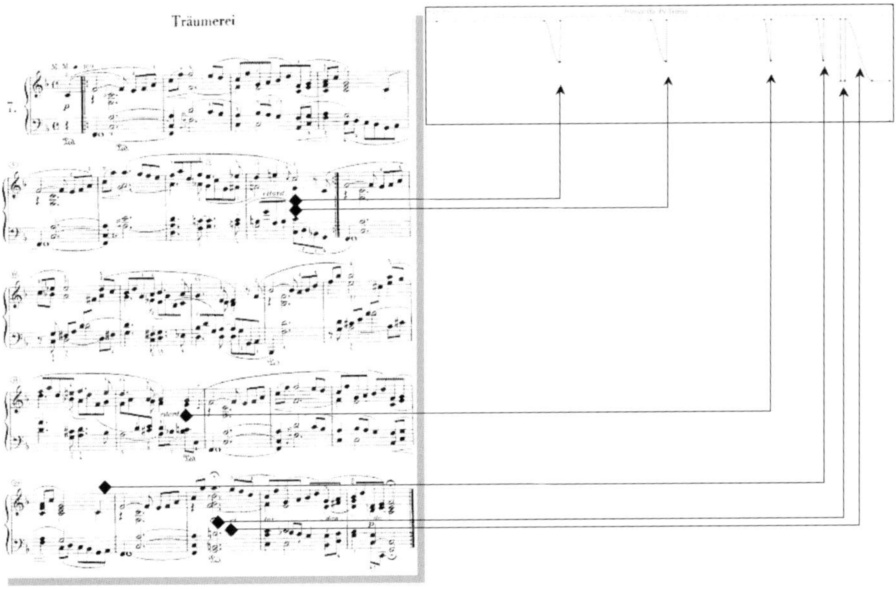

Fig. 16.9. The PrimavistaRubette deals with performance commands given from the score's structure. Here, we are giving the primavista agogics defined by the score's tempo indications.

The primavista weights are a special case, but one can interpret them in terms of weights. It deals with performance signs that are written on the score and need a representation by means of weights. We can do this for all primavista signs. Let us show how it is done just for agogics. In the example shown in figure 16.9, we have a set of tempo indications: several ritardandi and one fermata. This can be encoded as a weight function that shows a tempo curve that reflects these signs. The precise shape and quantity of these commands can be defined on the preferences of the PrimavistaRubette, so it is up to the user to make precise the meaning of a ritardando or a fermata. But the resulting curve is understood as a weight function, which, when applied to tempo shaping, yields the quantitative and qualitative forms of these agogical signs.

17

Shaping Operators

Those who are good at making shape don't usually fight.

Performance operators are those instances of our theory that shape a performance transformation. We have defined the relevant structures, namely performance cells, in chapter 10. A performance cell essentially includes the performance transformation $\wp$, the symbolic kernel K, and the sounding output data (the initial data are not central in this issue, so we neglect them here). Performace operators will also have to act on one of the three components of a performance cell: the symbolic kernel, the field defining the transformation $\wp$, or the physical sound output. According to one of these cases, we have called the operator *symbolic*, *field*, or *physical* (figure 15.10). Any performance operator will have to define such a performance cell.

This can be done in two ways: Either a completely new performance cell must be constructed or a given one is taken and then used to generate a new one. The first (uninteresting) case is known as *primavista performance*: One takes the score data and produces a primary performance with no artistically elaborate shaping. One could add the primavista operator as described in section 16.4, and that is all. That operator takes the weights and interprets them in a straightforward way. For example, it transforms the tempo weight function w_T as shown in figure 16.9 into a tempo curve without any change to the weight's shape. For example, if the default tempo is 100 [♩$/min$], then the primavista tempo is $T(E) = 100w_T(E)$. This produces a neutral first rendition.

As already pointed out in section 15.3, the second case is much more difficult and important since shaping operators must be applied in very different situations of performance with complicated conditions.

Before delving into delicate questions of this type, let us get off the ground with some easy operators. A very useful and easy symbolic operator is the split operator. It is used to split the given composition into parts that have to be treated separately, such as right and left hand for a piano composition, or pe-

G. Mazzola, *Musical Performance*, Computational Music Science,
DOI 10.1007/978-3-642-11838-8_17, © Springer-Verlag Berlin Heidelberg 2011

riods in order to shape such time slices independently. The operator takes certain parameters and creates cubes by defining intervals in those parameters. Then the composition (the symbolic kernel) is split into two portions: one being within the cube, and the complement. If we repeat this procedure, we may create quite sophisticated boxing configurations allowing for detailed processing. For example, it may be necessary to deal with a small motif or an ornament separately in its shaping. A trill, for example, might require a very special agogical treatment. It can also happen that we just need to redefine some symbolic objects for the sake of better symbolic representation. This might happen with regard to some time signatures or pitch shifting conditions, etc. This can be done with the symbolic operator. It allows for affine maps in any set of parameters. We omit the details here.

A second, relatively easy type of operator is the physical operator. It allows a weight to act upon any selection of parameters of the output of the given performance. This does not influence the transformation, nor does it change the symbolic kernel. It just takes the given output and then alters that data. For example, we may let a weight act upon loudness or duration or pitch, whichever.

We now want to give an instructive example of a tempo operator. Let us first deal with a straightforward idea to construct a tempo operator. We suppose that we are given a tempo field $T(E)$ in the performance cell that we want to modify by the tempo operator. The modification should be made using a weight function $w(E)$. The straightforward approach is to let this weight act as-is upon the tempo and to generate a new 'weighted' tempo $T_w(E) = w(E)T(E)$. This works supposing that the weight takes only positive values on the given time frame. Let us suppose this now. But how should we deal with the extension of this formula to articulation? If we are given a parallel field at the outset, we have $\partial T(E,D) = (T(E), 2T(E+D) - T(E))$. So we get the weighted parallel field

$$\partial T_w(E,D) = (T_w(E), 2T_w(E+D) - T_w(E)).$$

While this formula might work for a parallel field, if we apply it to a non-parallel field, it destroys the duration component of the given field and replaces it with the parallel component. This is precisely the delicate situation we alluded to above when saying that this is the straightforward approach: The given performance field might have a tempo component, but the articulation component (duration) is not a function of the tempo. Can such destructive action be avoided?

Yes, it can, and that works as follows. Restate the D component of the parallel field by

$$2T(E+D) = \partial T(E,D)_D + T(E)$$

and then get the formula for the weighted parallel field:

$$\begin{aligned}\partial T_w(E,D)_D &= w(E+D)(\partial T(E,D)_D + T(E)) - w(E)T(E)\\ &= w(E+D)\partial T(E,D)_D + (w(E+D) - w(E))T(E).\end{aligned}$$

So the total two-dimensional articulation field is as follows:

$$\partial T_w(E, D) = \begin{pmatrix} w(E) & 0 \\ w(E+D) - w(E) & w(E+D) \end{pmatrix} \partial T(E, D)$$

Call

$$Q_w(E, D) = \begin{pmatrix} w(E) & 0 \\ w(E+D) - w(E) & w(E+D) \end{pmatrix}$$

this matrix. Then we can define this field equation

$$\mathbf{Ts}_w(E, D) = Q_w(E, D)\mathbf{Ts}(E, D)$$

for an arbitrary articulation field. The definition is independent of $\mathbf{Ts}(E, D)$ being parallel or not. This is what can be taken as a generic definition of the tempo operator! Whenever we have an articulation field $\mathbf{Ts}$ on $\mathbb{R}^{ED}$, the tempo operator is just the one defined by the matrix Q_w deduced from the weight w with the above formula. Of course, this specializes to the weighted parallel field if the original articulation field is parallel. But it works in complete generality.

This example shows where lie the difficulties and subtleties in the construction of clever performance operators: They act in maximal generality upon given performances, but specialize to what is expected for classical special cases. This has a deeper meaning than just a technical flexibility. The entire operator theme is about what it means to conceive expressive performance. This concept is about how we insert rhetorical architecture into the performance's unfolding. What is it that we want to influence by a given weight, and what is the essence of such an influence? For example, the tempo operator: Do we really understand this operator if we just apply it to the parallel situation? The above construction shows that we can interpret the tempo processing by a weight as being a deformation of any given articulation hierarchy $\mathbf{Ts}(E, D) \to T(E)$.

Behind this concern for flexible operators is also the deep question about the variety of operators as such. How many operators do we have? Are they all essentially different or are there some generic operators that specialize to more specific forms? Is there even a unique master operator, which can be specialized to any specific type?

Why is this musically speaking relevant? Because we would like to know about the unifying principles of expressive performance, at least in the analytical domain.

17.1 Are Lie Type Operators Universal?

There is a type of shaping operator that is both a well-known construction in mathematics as it is a quite general approach in the musical context. It uses a classical operator in differential geometry: the Lie operator. The Lie operator is defined for a given vector field X on a performance frame, say. It takes a differentiable function f on that frame and creates a new function $L_X(f)$, the Lie derivative of f. This operator $L_X : F \to F$ on the algebra F of functions on the given frame acts as a derivation: It is $\mathbb{R}$-linear and we have $L_X(fg) = fL_X(g) + gL_X(f)$. The relevance of this construction lies in the fact that the map $X \mapsto L_X$ is an isomorphism of the vector space of vector fields onto the vector space of derivations. A vector field is essentially the same as a derivation, which transforms functions into functions. Therefore:

Performance fields are essentially derivations on weights. Which means that performance fields are naturally associated with weights.

This is a strong argument for both, the performance field formalism and the usage of weights for the shaping of performance. Let us define a general operator using a weight and acting upon a given performance field. Take a performance field $\mathbf{Ts}$ on the source space $\mathbb{R}^{X.}$ of a performance hierarchy. Let Z, S be two subspaces of the hierarchy, Λ a weight on Z, and $Dir : S \to S$ an affine endomorphism. Let $i_S : S \to \mathbb{R}^{X.}$ be the embedding map of S, and $p_S : \mathbb{R}^{X.} \to S$ the projection onto S. Then we have this new performance field:

$$\mathbf{Ts}_{\Lambda,Dir} = \mathbf{Ts} - L_{\mathbf{Ts}_Z}(\Lambda) i_S \circ Dir \circ p_S$$

with $\mathbf{Ts}_Z$ being the Z-component of $\mathbf{Ts}$. This shaping operator type is called a *Lie operator*. So the operator acts trivially if the gradient of Λ is orthogonal to the given field $\mathbf{Ts}_Z$, i.e. the integral curves of the performance field move along constant weight hypersurfaces. Which is completely natural: When moving along an integral curve, the weight does not change, so it should not affect the given performance.

The point of this Lie type operator is that it cover quite a number of operators. Namely all those that create one of the following three deformations of hierarchies:

- The articulation hierarchy $\partial T \to T$ deforms to $Z_w \to T_w$ for a given weight w by the above matricial operator Q_w.
- The parallel articulation hierarchy $\partial T \to T$ deforms to a general hierarchy $Z \to T$.
- The hierarchy

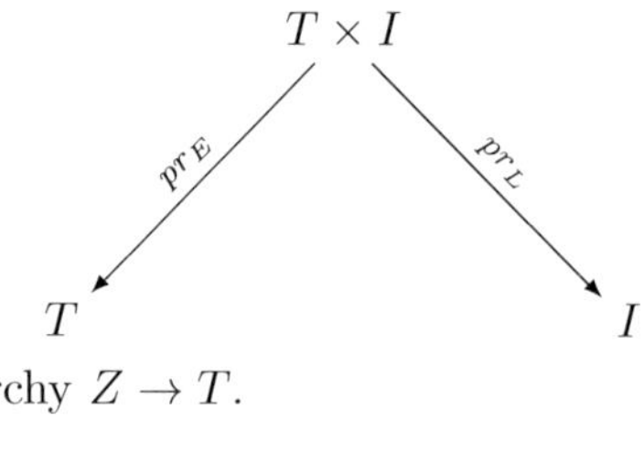

deforms to the hierarchy $Z \to T$.

18

Two Generic Models and the Challenge of Improvisation

Mon Dieu, donnez-moi de la simplicité!
Leo Tolstoi

Besides partial models with emotional, gestural, or analytical rationales, there are two models that comprise all these approaches. We want to briefly present them for completeness, and less because they offer a deeper insight into the complex of problems related to expressive performance.

Although this book is not about improvisation, we believe that it is important to give a short introduction to the perspectives that are opened when classical performance is extended to improvisatory approaches. What are the relevant differences? Can the concept or performance still be applied to improvisation, or do we have to open new conceptual spaces transcending or even negating the model developed so far by contemporary performance research? We address these questions in section 18.3 below and make a case study on Miles Davis' 1964 interpretation [22] of the classical jazz tune *I Thought About You*.

18.1 The GERM Model

The first model is the GERM model. It has been described in 2001 by Anders Friberg, Patrik N. Juslin, and Roberto Bresin [55]. The GERM acronym means

- G = Generative Rules, which function to convey the generative structure in a musical manner; these rules are those given by the Director Musices rules of the KTH system.
- E = Emotional Expression, which is governed by the performer's expressive intention; these were discussed in our discussion of emotion-based expressivity investigated by the KTH school, although the GERM emotion catalog is slightly extended.

G. Mazzola, *Musical Performance*, Computational Music Science,
DOI 10.1007/978-3-642-11838-8_18, © Springer-Verlag Berlin Heidelberg 2011

- R = Random Variations, which reflect internal timekeeper variance and motor delay variance; these factors stem from motorically and psychologically variable uncontrolled variations.
- M = Movement Principles, which prescribe that certain features of the performance are shaped in accordance with biological motion. This relates to the classical Sundberg-Kronman-Verillo-Friberg-Todd assumptions on retard behavior of tempo.

The model supposes that the above four components are relatively independent factors, although they might be coupled. The model has been implemented on Director Musices and tested empirically with different psychometrical tests.

18.2 Todd's Generic Approach

Neil McAgnus Todd's generic approach to semiotic expressivity in performance [136] runs as follows. His performance model is designed upon a bidirectional transformation pairing from a score representation Ψ to a performance P and backwards by means of:

1. a *performance procedure* Π acting on Ψ and an *encoding function* γ: $P = \Pi(\Psi, \gamma)$,
2. a *listening procedure* Λ acting on P and a *decoding function* δ: $\Psi = \Lambda(P, \delta)$.

In this generality, "the theory... is sufficiently general to cover any variable of expression. At the same time, it is agnostic as to what is being communicated, be it structure, emotion, or extra-musical reference" [136, p.407]. The generic character of Todd's approach hides an asymmetry of the transformation pairing, which is due to its semiotic background. In fact, performance is a poietic process issued by the performer from the composer's score. In other words, a performance is *caused* by its creators and must be *understood* by the listener, not vice versa. Hence, the performance transformation has to be specified as a semiotic mechanism. This is the difficult part of the business, and we have discussed this extensively in chapter 4.

The critical subject of performance theory—a problem which Todd thematizes in the spirit of cognitive science—is a reconstruction problem: Given a performance P, how many representations Ψ and encoding functions γ can you find such that $P = \Pi(\Psi, \gamma)$? In mathematical terms, we are looking for the fiber $\Pi^{-1}(P)$ over P. This is the so-called inverse image of P, and therefore, this branch of performance theory is called inverse performance theory; we shall come back to this topic in chapter 24. The listening procedure in [136] is just a formal setup for a section Λ to Π, i.e., the selection of an element in the fiber over P as a function of the decoding data δ.

Clearly, the fiber cannot be described in effective mathematical terms if one does not assume a well-defined transformation model. And even for very special models, the so-called locally linear performance grammars [81], fibers turn out to be high-dimensional algebraic varieties. Further, the encoding function must be meaningful enough to reflect the score's structure and its relationship to expressive semantics. Otherwise, performance cannot claim to interpret the selected score. In other words, the big problem of performance theory is to propose models of *adequate generality* that cope with *semiotic expressivity*.

In Todd's singular example to his theory, he restricts to hierarchical grouping data for the shaping of duration. Commenting on the inverse problem of listening procedure, he states: "The durations used in the calculations are from only one metrical level. Much information about tempo is given at metrical levels below the tactus and in the durations of actual notes. The representation needs to be extended downwards to include note timing, which would mean that a rubato handler would have to work in cooperation with a metrical parser, one feeding the other. Clearly, a lot of work is needed in this area." Concluding, he notes: "The known algorithms make no reference to any tonal function. Therefore, a rubato handler could be a vital component of any theory of grouping in the perception of atonal music. A complete theory must of course include dynamics, articulation and timbre."

Methodologically, this approach is tightly bound to cognitive science in that any algorithm is first of all tested upon its immediate fitting into human perception mechanisms, within real-time constraints, say. We believe this is a too-narrow approach for two reasons. First of all, the investigation of general structural facts must be carried out before any relevance to human perception is taken into account. There is the general problem of getting an overview of possible models and their classification. Second, the cognitive knowledge is all but settled. More precisely: We do not know the processes by which cognition of performative expression is handled in the human brain. It could happen that a rather abstract invariant of the geometric structure of a mathematically complex fiber $\Pi^{-1}(P)$ can easily be detected by the cognitive machinery, but that this invariant would not have been detected if we were only permitting fibers that allow an immediate access by the cognitive capacities. For example, the mathematical structure of a Möbius-strip-shaped fiber may be too complex to be grasped by the cognitive machinery, whereas its lack of orientation may be an easy task to be tackled by a small test routine built on a neuronal basis.

18.3 The Challenge of Improvisation

This book is not about improvisation, but we should nevertheless embark in a short discussion of why the present approach to performance theory would miss improvised music making, and what this failure looks like in detail. There are many reasons to exclude improvisation from performance theory. We shall discuss some of the most evident theoretical arguments and then also illustrate them in an analysis of a prototypical improvisation, namely Miles Davis' March

1964 recording (♫ 18) of Jimmy Van Heusen's popular 1939 song *I Thought About You* (lyrics by Johnny Mercer).

The nature of lead-sheet-driven improvisation in jazz derives from the fact that the lead sheet is a symbolic score that will not be performed as is, but must undergo a series of delicate symbolic transformations before the sounds emerge from the improvisers' playing. The question here is how such a transformational process can be described. To begin with, it fits perfectly in our model of multi-agent communication described in section 4.11. The composers' field is no longer occupied by the classical score composer but results from a multi-layered activity that eventually leads to the production of sounds. This improvisational process is far from amorphous sponteneity however. The lead sheet, with its reduced symbols comprising the basic melodic shapes and the metrically displayd chord symbols that drive the improviser's harmonic changes, is enriched in a quite logical layering of improvisational spaces with a sucessively richer ambient structure of strategies and actions.

When we look at those layers, we recognize as input the mechanical data of the lead sheet and then as output the notes, which will be played on the instrument to yield the improvisation's sounds. The entire transformation process is a symbolic one in the sense that the improviser constructs the output on a mental level. This construction is well known among jazz musicians and is called the "inner score" ("partion intérieure," also the title of an excellent jazz theory book [128]). We shall see in the subsequent analysis of Miles Davis' 1964 recording of *I Thought About You* that the nature of this symbolic activity is however not executed as an algorithm on the level of note symbols. These symbols are given, but their creation lives in the stratum of gestural embodiment. In other words: The improvisational process is a complex gestural activity. It relates to the factorization of performance as discussed in section 4.6 and illustrated in figure 4.6.

The immediate question arising from this approach relates to the space where such gestural activity takes place. We have given a first approximation to this question in section 4.12. The improviser's creative space is that realm of artistic presence that we called "imaginary time-space." The artist's activity is a gestural movement, a dancing gesture that acts upon the given lead sheet symbols and moves them into the final output according to well-defined strategies of embodiment. If one had to insert this creative activity into our scheme of shaping operators, it would be a system of symbolic operators (a small set of elementary symbolic operators is implemented in the PerformanceRubette of Rubato, see section 15.3).

The gestural character of this process becomes evident from the movement that the improviser constantly applies to the given material in his/her shaping of the played notes. The output resembles an embryonic growth movement of a living body; it is the body of time that is being shaped and whose dynamic anatomy is being created. In what follows, we shall learn about this improvisational realm, which we call *improvisational time-spaces.* From the above reflections, it follows that these time-spaces live in the imaginary time-space.

And it follows that such an improvisational style is everything but blind spontaneity. It is a highly developed cultural code within embodied thinking in the flow of creation.

18.3.1 Expansion of Improvisational Time-Spaces

Lead-sheet-driven jazz improvisation is most often governed by a basic melodic line, harmonic structure of chord changes, and metered time. The concept of time can be understood from the combination of tempo and meter with emphasis of strong and weak beats, but, as we shall see, extends to a more in-depth phenomenon when shaped by the creative improviser. Most Jazz standards contain regulatory phrases of four to eight measures in length and are derived from popular songs. This is also true about Jimmy Van Heusen's tune *I Thought About You* and will be the basis of the following analysis of the expansion and evolution of improvisational time-spaces; see figure 18.1 for Lisa Rhoades' transcription of the 1964 recording. The lower staff shows the original melody, the upper staff shows Miles Davis' performance.

Traditionally, an instrumentalist finds *improvisational space* in a jazz standard where the melody is at a resting point. But we shall describe the improvisational process in more detail, strongly including shaping of time, since Miles Davis' performance unfolds way beyond the traditional space.

Fig. 18.1. Lisa Rhoades' transcript of Miles Davis' 1964 performance of the jazz tune *I Thought About You*. See text for details.

The conceptual synthesis of time and improvisational space into one term is specific to the type of improvisational development that we will examine. We shall screen this synthesis as a layering of four time-spaces: 1. Metronomic Time-Space, 2. Emergence Time-Space, 3. Progression Time-Space, 4.

Advancement Time-Space. In Rhoades' transcription, figure 18.1, the instanciation of such time-spaces is indicated by circled numbers ①, ②, ③, ④.

① **Metronomic Time-Space**
The first time-space is the mechanically driven or unvaryingly regular space in rhythmic and harmonic patterns. It consists of the formal background and structures of the tune as defined by the lead sheet.

② **Emergence Time-Space**
The second time-space is identified as a direct extension of the metronomic time-space. In our tune, this can be described by the use of a tritone substitution (the shift $T^6(X)$) sharing the same 3rd and 7th of the original seventh chord X, albeit reversed; for example $X = C^7 = \{C, E, G, B_\flat\}$, and substitution $T^6(X) = G^7_\flat$. So time is still the original one in this layer, only harmony moves on an extended gesture.

③ **Progression Time-Space**
In this time-space, time is reshaped: In our tune, we observe the emergence of a new meter, such as 6/8 over 4/4 in measure 4, by the extension of a triplet figure inherent in the original. This space creates tension and propels the improvisation to open a new time-space of rhythmical progression. The ensemble seems to be playing "out of time" in this third space, which we will therefore refer to as the Progression Time-Space. Here, the gestural flip-flop of micro and macro meter that the musicians play and react to causes a temporal shift. The acute listener can still determine the regular metronomic beats, but it becomes far more distant and is less important in the music making.

④ **Advancement Time-Space**
The fourth time-space is referred to as the Advancement Time-Space. It is created from the Progression Time-Space when the entire ensemble has broken away from the Metronomic Time-Space through the tension created by a number of temporal shift gestures. The Advancement Space can be described by attaching to every event in the new position (as played by Miles Davis) an arrow relative in direction and size to the original position. The fourth space's temporal shift causes the tension/force of the original, and the arrows represent the distance and direction, like a deformation of a "rubber strip of time." This one has variable "temporal elasticity," enforcing variable tensions that are reacting to the strength of the original metric positioning of the note(s) played.
In the first measure, Davis' trumpet states the melody, and he stretches time easily, because there is no rhythm section player behind him delivering a metronomic beat. Also, the melody's formal rhythm of a quarter note triplet is a rhythmic motif that leads itself to a spatial interpretation. On beat 2 in measure 11 and measure 12, Davis achieves the Advancement Time-Space through the use of quarter tone lifts and bends that flow above a predetermined rhythmical space. The space is thus extended through the next measure by the same lifting of tonal integrity and "breath."

In examining the traditions and innovations of jazz music, a musician will gain insight and pinpoint key areas for further improvisational development. Before we do the analysis of the tune, we should briefly comment on jazz education today, since this strongly relates to the work to be performed in the above time-spaces. Some examples of methods and strategies of jazz education used in collegiate instruction are Jamey Aebersold's play-alongs, playing by ear, transcription of jazz solos, and the "etude method" of distributing favorite jazz lines over chords in tunes and freely improvising around them; see [3] for the first of more than 128 volumes (!) of Aebersold's library. These methods focus on building a linear language of jazz and are quite successful. However, for many players that gain a wealth of jazz language and memorized riffs, it happens that their own voice remains vacuous and they lack the innovative inspiration of improvisation that is the essence of jazz expression. But innovation is most often made possible by musicians deeply rooted in their tradition and having their own identity within the tradition that enables them to absorb its secrets and create something new. Through research and analysis of jazz traditions and innovations, as presented in our transcription and analysis of the expansion of improvisational time-spaces in Miles Davis' recording, we hope to help jazz improvisers evolve and deepen their understanding of the high culture of improvisation. Through the analysis of our transcription of Davis' performance, the musician can discover expansive improvisational time-spaces and ingenious use of linear and harmonic nuances.

18.3.2 The Analysis

The tune *I Thought About You* is in ABAB form with a lovely melody based on a descending half step within a series of ii, V, I's (e.g. measures 2-3, 8-9, 15-16). The A sections are identical and the B sections are varied with the B section ending on the tonic F. Our transcription only includes ABA because in Davis' performance, the second B section is in the beginning of his solo.

Like Dinah Washington in her recording of the song in 1954, Davis creates a bluesy ambience of a classic "torch song," a sentimental love song in a setting of a nightclub at 2 a.m. In the third measure of the tune (②), Davis plays an $A_\flat$ which is the minor 3rd, blue note, of the song's F major tonality. The ingenious quality of this note choice is that the G^7 chord is still sounding, then chromatically resolves up to $A_\flat^7$. The listener hears the halfstep clash but naturally relates it to the melodic phrase in F major as a blue note. Davis states the melody as a vocal line of a torch song by his rhythmic displacement of almost every entrance of a phrase except for the beginning of the first measure and beat 4 of measure 12. This allows his lines to breathe and swoon in a sentimental fashion that is a traditional characteristic of a torch song. The high point of note E of the first phrase (measure 1, ③) is rhythmically displaced and introduces a new improvisational time-space on beat 4. The result of this new time-space creates anticipation in the listener and room for new possible improvisatory development.

In the beginning of measure 2 the Advancement Time-Space is created by hinting at the next original phrase that begins on $B\flat$ on beat 4. Davis achieves this new-time space by playing the $B\flat$ as a grace note on beat 1. In measure 3, Davis plays a rhythmic diminution of triplets of the original phrase ($B\flat-A-F$) that begins on beat 3 of measure 2. The beat 4 rhythmic placement of the triplet figure emphasizes the "swing" feel.

In measure 5, Davis introduces a new time-space again with rhythmic displacement and diminution. This time Davis delays the entrance of the phrase and masks the melodic outline by traditional bebop ornamentation, which is to approach or leave the goal note by a half step and surround the next goal note, creating what is called a *turn*. This ornamentation most always resolves to the second goal note from a whole or half step above. Davis also extends the melodic line up to a high F at the end of measure 5 and extends the harmonic chord structure of the F^7, which sounds on beat 4 of measure 5. Davis outlines the Fm^7 chord by descent to resolve on the melodic note A.

The extremity of diminution and the hint of bebop in this line propels the rhythm section to increase in speed. The entire ensemble plays together on beat 1 of measure 7 and naturally relaxes with Davis playing a half step melodic motif that is played in half notes and restated in quarter notes ending on the augmented 4th scale degree of a Cm^7 chord. It should be noted that the augmented 4th scale degree is particularly characteristic of Daviss style of playing. Playing a sharp 4th scale degree over any given chord allows it to be heard as a chord tone, whereas the natural 4th scale degree will be heard as a wrong note. Also, note that hearing the 4th scale degree as a chord tone expands the linear structure of the original scale and adds a new sonority. This is where a player can delve deeper into this new harmonic space by playing a tritone substitution. This completes the initial analysis of the A section.

The first phrase motif of the B section begins with a rhythmic displacement of a sixteenth note and is tastefully stated in diminution, ending on beat 2. The original phrase motif begins on the & of beat 1 in measure 9 and ends on beat 3. Miles introduces "swing" feel in measure 10 with triplets, where the original is stated by a quarter note followed by two eighths and another quarter note. Davis continues to introduce new improvisatory time-spaces (③) by these triplets as part of the same syncopated feel of his first phrase of the B section and the next phrase.

Davis cleverly states the melody as a transposition of a fifth above the original in measure 11. He begins by outlining the D^7 chord and then ascends to his transposed quotation of the melody. The high G is the 7th of the $A^7_\flat$ chord on beat 3 of measure 11. The 7th and 3rd scale degrees are the two scale degrees that define the chord's quality. Finally, Davis places beat 4 in the same rhythmic position as the original melody in measure 12. He then deviates from that rhythmic unity by scooping up to beat 1 of measure 13 and descending in quarter notes followed by two triplet figures. Davis states the original melodic phrase of measure 11 in measure 15 in triplets. In measure 16, Davis rises from a scoop of the note A to B and descends chromatically to note G of the melodic line. This technique introduces a new harmonic space and time-space (②).

Miles plays his highest note once again as an extension of the restatement of the initial melody in the A section. Without notice, Davis erupts with a double high F in double-time feel to complete the last four bars of the first phrase. This high F is followed by an arpeggiation of descending quarter notes and repeated low note Fs in syncopated triplets. The listener might expect Davis to rest on low F, but instead he energizes them with his syncopated eighth note triplets. It must be noted that Davis deliberately deviates from the traditional form and begins his solo in the last four measures of the last A section. The listener traditionally would listen to another B section of repeated melody as well. This deliberate move to begin soloing before the melody is traditionally completed suggests another new time-space (④).

In Marshall Bowden's review of the Miles Davis CD set *The Complete Miles Davis at Montreux* [7], he states, "The 1960s quintet featuring Herbie Hancock, Ron Carter, Wayne Shorter, and Tony Williams had taken traditional jazz forms to as abstract a place as was possible." Davis was searching for a new stream to take his music and get away from the traditional theme/improvisation/theme scheme of jazz. The tune *I thought About You* is a pivotal example of Miles Davis' vision and concept of melody, rhythm, and form.

19

String Quartet Theory

Bei der nämlichen Gelegenheit fragte ich Haydn, warum er nie ein Violinquintett geschrieben habe, und erhielt die lakonische Antwort, er habe mit vier Stimmen genug gehabt.[1]
Ferdinand Ries [116, p.287]

Fig. 19.1. Ludwig Finscher.

Why are we including this topic in a book on performance theory? Couldn't one include any other topic relating to a specific musical genre as well? Solo piano music, operas, what not? The reason for the special role of the string quartet is that this genre has a theory, a reflection about why these four instruments—the two violins, the viola and the violoncello—merge to such a perfect harmony or collaborative music, a theory that is not only relevant sociologically, but also with respect to the intrinsic messaging of musical contents.

But there is more: We have a string quartet theory in the sense of musicology or music theory as prominently described by Ludwig Finscher in his habilitation [32]. And we have a mathematical theory of the string quartet as developed by the author and presented in [75] and reproduced in [84]. This second theoretical approach perfectly confirms Finscher's findings about the singular role of the string quartet in the European history of music. But it is more prominently also a theory of the role of musical instrumental parameters in their function as expressive tools for the communication of musical contents. In this sense this theory is about performance: It deals with the question of how analytical musical contents are conveyed to the audience in the exquisite instrumentalization of the string quartet.

[1] On the said occasion I asked Haydn why he had never written a violin quintet, and I obtained the laconic answer that he had always had enough with four voices.

G. Mazzola, *Musical Performance*, Computational Music Science,
DOI 10.1007/978-3-642-11838-8_19, © Springer-Verlag Berlin Heidelberg 2011

The general topic behind this chapter is still another: It deals with the general question about the adequacy of instrumentation with respect to the musical contents of a composition. Why do we need a certain instrumentation to express a musical thought? Would it work if we played Beethoven's op.130 with four recorders? Or would it be advantageous to play *Happy Birthday to You* by a Wagner orchestra? The latter would be 'overdressed' while the former would flatten down the musical depth. It is interesting that this type of question has seldom been dealt with in classical orchestration and composition.

19.1 Historical and Theoretical Prerequisites

Finscher describes the historical root of the string quartet as follows (translation by the author in [84, p. 994]):

> *The prehistory of the string quartet is more complicated than that of any other instrumental art form of the eighteenth century. It cannot be causally deduced from any single one of the threads of tradition from where it comes.*
>
> *To a certain degree it is the creative act, the invention out of a moment of the delicate historic equilibrium, the kairos in the sense of ancient Greek thinking.*
>
> *The prehistory dates only from about 1720 to 1760 when Luigi Boccherini and Joseph Haydn independently invented the string quartet. In 1761, Boccherini wrote his first quartets in northern Italy, they were published 1767-68 in Paris under the name of "quatuor concertant." Probably Haydn had written quartet "divertimenti" already in the 1750s in Vienna, they were however only well known in 1760.*
>
> *The sparse regional, instrumental, and stylistic rootedness in the string quartet's prehistory, from which this new art form has quite spontaneously emerged, provokes the question whether beyond historical rationales a more systematic understanding could better enlighten the 'string quartet phenomenon.' The problem is to question this precise date (1760) of the rise of this precise instrumental art form (the string quartet) in the context of the European music from the systematic point of view.*

We should recall here the famous words of Carl Dahlhaus [20, vol. 10, p.104-105]:

> *Erst die systematische Konstruktion öffnet den Blick dafür, welche Tatsache einer Geschichte angehören, die zu erzählen lohnend erscheint.*

Only the systematic construction is capable of shedding light upon those facts of a story that are worth being narrated. The string quartet is a particularly challenging historical fact. Why did it emerge so unattendedly around 1750? And then grow instantly to a most profiled genre? Finscher:

> *Das Streichquartett ist die einzige Gattung der neueren Instrumentalmusik, die eine solche an einem einzigen künstlerischen Modell entwickelte, vergleichsweise genau und detailliert ausformulierte und als allgemeinverbindlich akzeptierte Theorie ausgebildet hat.*

The string quartet is the only genre of more recent instrumental music that has developed a comparably detailed and generally accepted theory built upon a single artistic model. Therefore, it is a good point of departure for our performance-oriented discussion of this genre that we have a good musicological theory. The model is based on

- the four-part texture,
- the topos of a conversation of four humanistically educated persons,
- the distinguished character of the family of violins.

The four-part texture ([84, p. 995])

The four-part texture was the ideal type of structured polyphony that was oriented on the counterpoint with its long tradition. This is the formal—or better: formalized—element of string quartet theory. We have to take it in the full conceptual ambiguity, i.e., on the one hand, the texture is "note against note" in its linear temporal progression in the sense of classical counterpoint. On the other, it is a texture of vertical units as an expression of harmonic relations.

With Haydn, one could imagine that he could have added a fifth voice to "enrich the texture." But it is reported that he 'failed' on several occasions with this 'experiment.' Ries reports 1838 ([32, p.287]):

> *Bei der nämlichen Gelegenheit fragte ich Haydn, warum er nie ein Violinquintett geschrieben habe, und erhielt die lakonische Antwort, er habe immer mit vier Stimmen genug gehabt. Man hatte mir nämlich gesagt, es seien drei Quintette von Haydn begehrt worden, die er aber nie hätte komponieren können, weil er sich in den Quartettstil so hineingeschrieben habe, dass er die fünfte Stimme nicht finden könne; er habe angefangen, es sei aber aus einem Versuche am Ende ein Quartett, aus dem anderen eine Sonate geworden.*[2]

The topos of a conversation of four humanistically educated persons ([84, p. 996])

This topos must be cautiously distinguished from the well-known topos of a "sound speech," i.e. from the similarity of musical expression or semantics to the common language. In the case of the string quartet, what's more important

[2] On the said occasion I asked Haydn why he had never written a violin quintet, and I obtained the laconic answer that he had always had enough with four voices. They had told me that three quintets had been commissioned to Haydn, which he however was never able to compose because he had become so accustomed to the quartet style that he could not find a fifth voice; he had commenced, but from one attempt a quartet emerged, and from another a sonata.

than speaking is the dialogue, a fact that becomes more evident in the French expression "quatuor dialogué" for "string quartet" (in fact the invention of a publisher). The association of a discourse to the string quartet was initiated by the musician Johann Friedrich Reichhardt in 1777 ([32, p.287]):

> *Bei dem Quartett habe ich die Idee eines Gesprächs unter vier Personen gehabt.*[3]

Like Haydn, Reichhardt also views the number four as being the upper limit for a good dialogue. He (also) tries to add a fifth person to the quartet. But he fails:

> *Die fünfte Person ist hier ebensowenig zur Mannigfaltigkeit des Gesprächs nothwendig, als zur Vollstimmigkeit der Harmonie; und in jenem verwirrt sie nur und bringt Undeutlichkeiten in's Stück.*[4]

The same happens to Schumann ([32, p.289]) in a discussion about a viola that was added in a quintet:

> *Man sollte kaum glauben, wie die einzige hinzugekommene Bratsche die Wirkung der Saiteninstrumente auf einmal verändert, wie der Charakter des Quintetts ein ganz anderer ist, als der des Quartetts. Hat man im Quartett vier einzelne Menschen gehört, so glaubt man jetzt eine Versammlung vor sich zu haben.*[5]

The distinguished character of the family of violins ([84, p. 997]) The formation of the string quartet would not have been thinkable without the collaboration of the homogeneous sound of the instruments of the violin family.

Finscher describes ([32, p.124/125]) the characteristic of violins as compared to the gambas, which were the preferred solo instruments in the seventeenth century, as follows:

> *Die Violinen hatten gegenüber den Gamben jedoch noch eine weitere zukunftsträchtige Eigenart: Sie gliederten den Tonraum, der ehemals in Analogie zu den menschlichen Stimmengattungen gebildet, nun aber in der Tiefe wie in der Höhe längst kräftig erweitert worden war, klarer und sinnfälliger, mit deutlicherer Individualisierung ihrer jeweiligen Tonbereiche. (...) Für das klassische Streichquartett, das die Beweglichkeit, den Lagenwechsel, den Kontrast- und Farbreichtum des symphonischen Streichersatzes mit der grösstmöglichen Annäherung an eine*

[3] For the quartet I had the idea of a conversation among four persons.

[4] Here the fifth person is as superfluous for the manifold of the conversation as it is for the full voicedness of harmony; and in the former it only creates confusion and makes the piece cloudy.

[5] It is incredible how a single added viola alters the effect of the string instruments, changing the character of the quintet radically as opposed to the quartet. If you heard four single persons in the quartet, you believe that now you are facing an assembly.

streng auskomponierte Vierstimmigkeit zu verbinden suchte, bot sich das vierstimmige Ensemble aus Gliedern der Violinfamilie als das ideale Instrument an.[6]

19.2 The General Plan

The general plan of our performance strategy of the string quartet is to generate optimal rhetorical expression of contents by use of the violin family's sound variety. This means that the sound configurations of the composition with respect to their constituting structures derived from counterpoint and harmony in the late eighteenth century must be played with maximal profile. What does "profile" mean? We must represent these configurations aurally such that these sound objects are optimally distributed in the sonic space-time. Despite this somewhat generic requirement, there is one condition that seems to be mandatory for optimal distribution: independent position, namely the condition that no three sound objects be aligned on a line, no four of them be lying on a plane instead of spanning a tetrahedron, and so on: any $n+1$ different points span an n-dimensional space. This is known as *general position*.

At first this looks more geometric than musical, but it was music theory that first introduced this seemingly geometric idea with the basic intervals of an octave, fifth, and major third in their just tuning. It is, in fact, well known that the pitches of all usual tunings (just, mixed, tempered) are (up to a basic reference pitch, such as the chamber pitch) rational linear combinations $p(o,f,t) = o\log(2) + f\log(3/2) + t\log(5/4), o, f, t \in \mathbb{Q}$. The three basic directions of these linear combinations, $\log(2), \log(3/2), \log(5/4)$, are linearly independent over the rationals; they are vectors of independent space directions. This was exactly how music teorists and composers conceived these basic intervals: They define musically independent harmonic "dimensions." That this is true in a very precise sense of modern linear algebra is a remarkable fact, and all the more since the mathematical machinery was invented several centuries after music's discovery.

In this spirit, general position will be the target to address when playing note assemblies with those rich instrumental colors. We therefore first have to make ourselves knowledgeable about the space of sounds that is opened by the violin family. When we know about that, we may ask questions about the

[6] Compared with the gambas, the violins had a further seminal peculiarity: They articulated the sound space, which originally was built in analogy to the human voices, now being however strongly extended in low and high pitches, with more clarity and evidence, with pronounced individualization of the respective regions of sound. (...) For the classical string quartet, which attempted to ideally connect flexibility, change of position, richness of contrast and colors of a symphonic string setting with the optimal approximation of a complete four-voice setting, the ensemble built from the four instruments of the violin family offered the ideal instrument.

configurations of sound objects in general positions, which are possible in such a sound space.

When we have those answers, we shall finally come back to the contents to be conveyed by these rhetorical sound representation tools. These contents must be analyzed with respect to general positions and then yield the number of instruments necessary to enable such rich-sounding representations. The result will be remarkable, but let us first get off the ground with the analysis of the violin family's sounds.

19.2.1 General Position for Performance: Four Examples

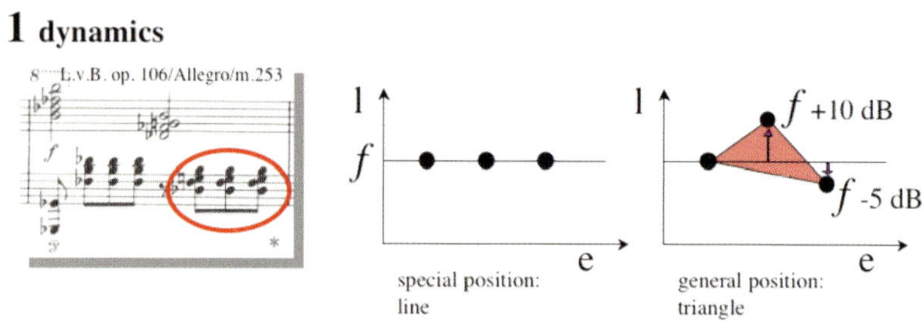

Fig. 19.2. Dynamical shaping of three chords in special position.

Before embarking on the systematic analysis of the sound space of the violin family, we want to discuss four examples of putting sounds in general position for performance purposes.

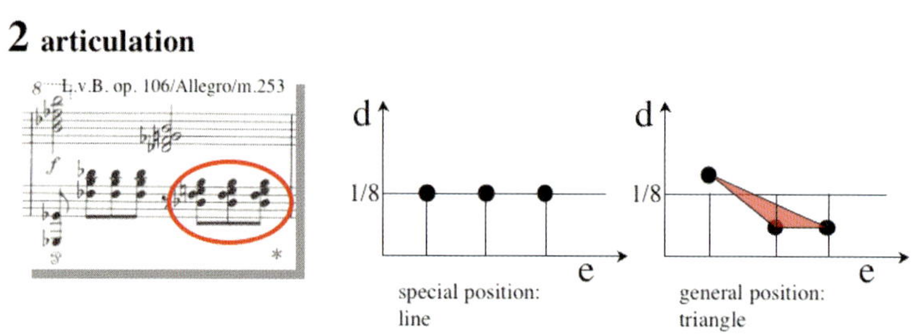

Fig. 19.3. Articulation shaping of three chords in special position.

The first example deals with dynamics (figure 19.2). We are given a piece where we want to express the rhythmical role of a sequence of chords in the left hand. They are all of identical loudness and occur an eighth note apart from each other. So they are in special position in the plane of onset and loudness. To change this, we increase loudness of the second by 10 dB and lower loudness of the third chord by 5 dB. The result are three onset loudness points in general position—they define a proper triangle.

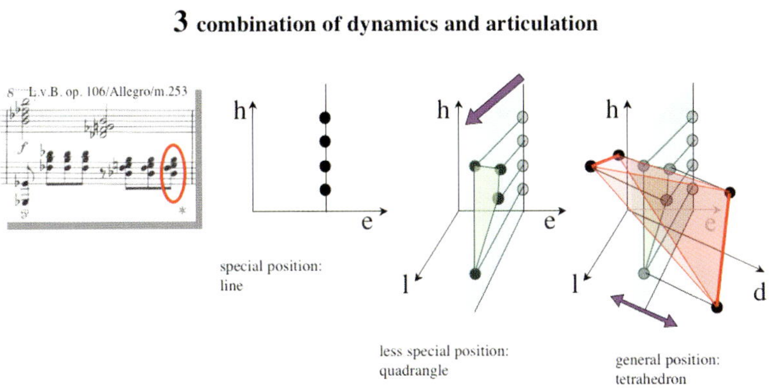

Fig. 19.4. Dynamics and articulation shaping of four chord notes in special position.

The second example is the same, but we now differentiate articulation instead of dynamics (figure 19.3). So we work in the plane of onset and duration. We are shortening the second and third notes, while the first is extended. The onset of the second is also shifted to enable the longer first duration. So the second starts while the first is still playing, and then we hear two staccati.

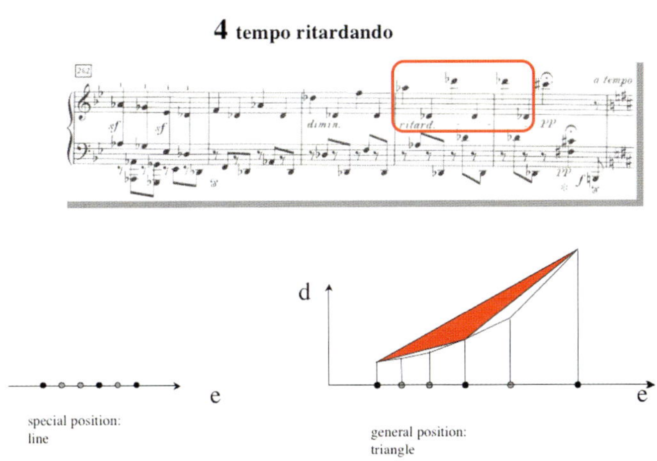

Fig. 19.5. Dynamics and articulation shaping of four chord notes in special position.

In the case of figure 19.4, four notes are all in one line on the pitch axis and have identical loudness, duration, and onset. To put them into general position, we first alter their loudness: The lowest and the highest notes are made louder than the two internal notes of the chord. The notes are now somehow more generally positioned: They build a quadrangle. The second step increases the duration of the lowest and the second to highest notes, whereas the other two notes are given shorter duration. The result is a tetrahedral configuration. Now the four notes are in general position.

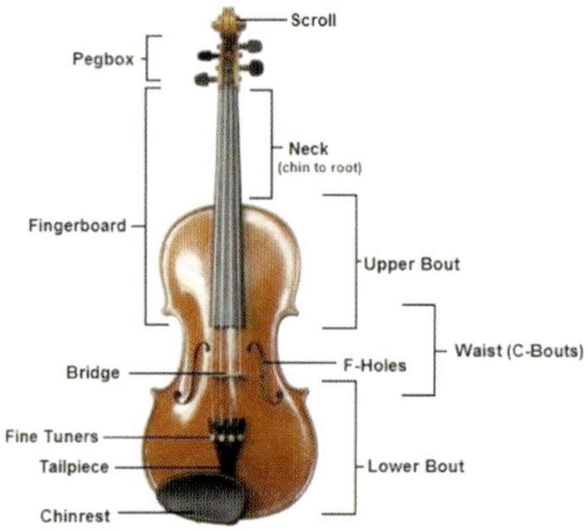

Fig. 19.6. The violin's components.

The fourth example, figure 19.5, shows the general position effect induced by a non-linear ritardando. We see that in the plane of onset and duration, the six points are such that some of the three-element subsets are in general position—for example the first, fourth, and last. To put the entire series of six notes into general position, we need a five-dimensional space. This is possible if we consider the parameters of onset, pitch, duration, loudness, and sound color (supposing that this one can be shaped here).

19.3 The Sound Space of the Violin Family

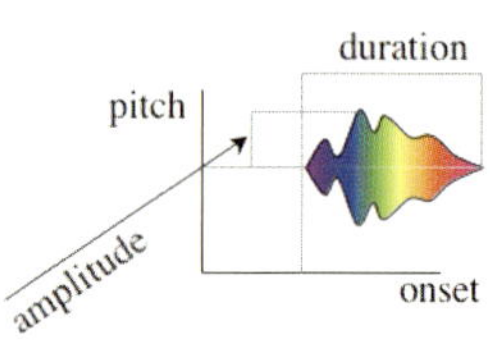

Fig. 19.7. A sound with its four geometric parameters.

The sound space of the violin family is a complex one, so let us start with the simplest information about a violin's anatomy (figure 19.6). The violin has a number of loci where it can be played, and this can be done in different ways, so, for example, the contact point of the string, or the angle of the bow on the string can vary. Let us give the complete list of parameters and then discuss them:

1 *Geometric parameters*
   a) onset, duration, and pitch define a three-dimensional space G (loudness will be added via amplitude to the technical parameters)
2 *Sound color parameters*
   a) instrumental parameters:
   violin type, choice of strings, performance conditions
   b) global technical parameters:
   vibrato: delay/pitch modulation (range of finger movement)/ modulation frequency/
   amplitude modulation (contact point of finger tip)

bow angle, bow contact point
they are a strong function of the individual player and remain relatively stable in time;

c) local technical parameters:
bow pressure, bow velocity = two-dimensional space H
they can be steered quickly and independently of each other, while bow pressure relates to amplitude.

The characteristic feature of the violin family is that the instrumental and the global technical parameters enable a much larger variety of sound color vectors than other instrument families. For example, if we compare two pitches, g and $g_{\#}$, being played by a Guarneri and by an F-horn, we get the image shown in figure 19.8. The Fourier spectra of the two instruments for the two pitches are markedly more independent from each other with the Guarneri violin as compared with the F-horn.

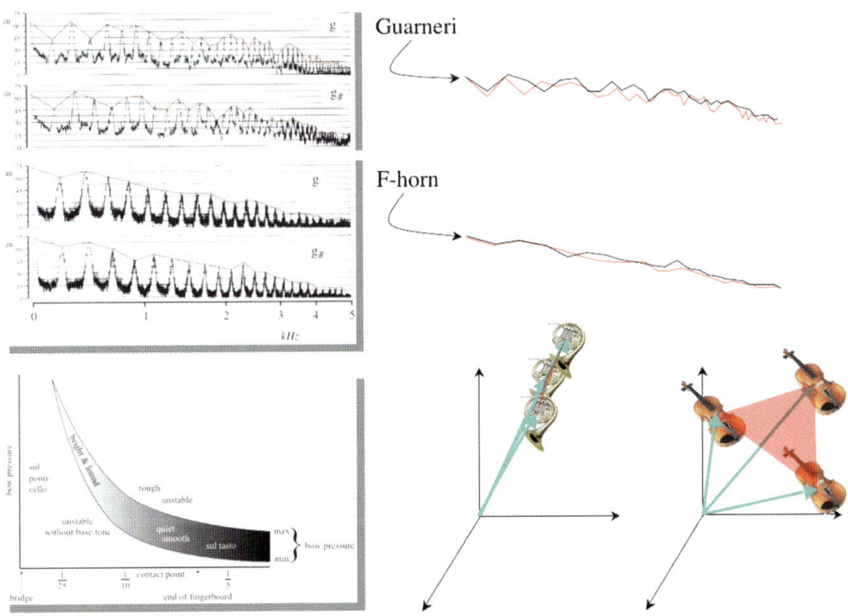

Fig. 19.8. The Fourier spectra of a Guarneri violin for two pitches g and $g_{\#}$, as compared with that of a F-horn. In their representation as vectors, these spectra look much more independent for Guarneri. The lower left graphic shows the variety of ways to play a violin.

19.4 Notes in General Position

The above sound space for n instruments of the violin family can be decomposed as follows:

1 Given n instruments of the violin family, we have a sequence $v_1, v_2, \ldots v_n$ of linearly independent vectors defined by the instrumental and the global technical parameters;
2 the two-dimensional space H7index$H$ of local technical parameters: bow pressure and bow velocity
3 the three-dimensional space G of geometric parameters: onset, duration, pitch.

This configuration enables the placement of sound events as played by the n instruments in a big space, see figure 19.9.

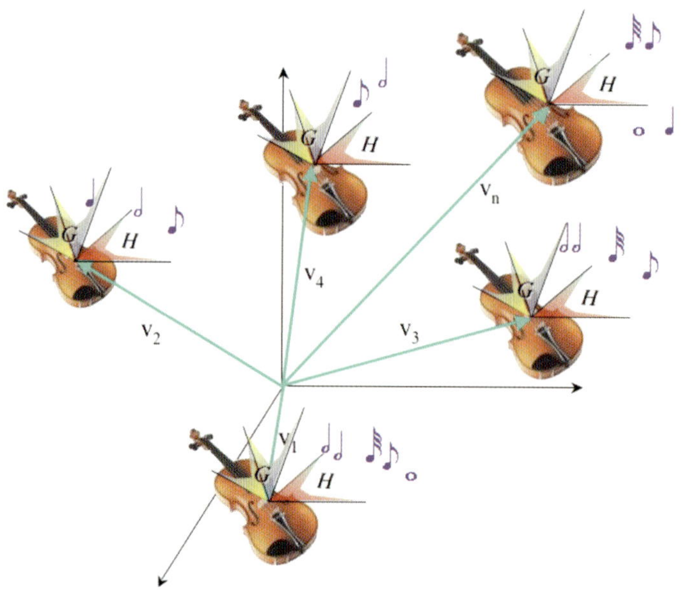

Fig. 19.9. The sound space of the violin family enables a distribution of the sound events played by n such instruments in a big sound parameter space.

As we are interested in general position questions, we need the answer to the question about how many sound events can be maximally placed in the space spanned by the instrumental vectors at which the common spaces G, H are attached, i.e. in the union

$$U = \bigcup_{i=1,\ldots n} v_i + G + H.$$

The answer is this mathematical theorem:

Theorem 1 *With the hypotheses and notations made on the parameter spaces available for the instrumental variety of the violin family, an ensemble of n such instruments can play at most $n+5$ points in general position on the space U.*

This means that when we look at all notes being played within a given string quintet, i.e. $n = 5$, then it is not possible to select more than 10 notes in general position.

19.5 Performance of the String Quartet

We now have the general geometric conditions for the display of notes in general position. But we have to add the music-theoretical situation given from the historical moment of the ending of the eighteenth century in order to understand the requirements for general position of notes as an expression of those theories.

For the contrapuntal theory, the essential structure is the first species. It is the basis for all other species, and we want therefore to look at the basic configuration of notes in this theory.

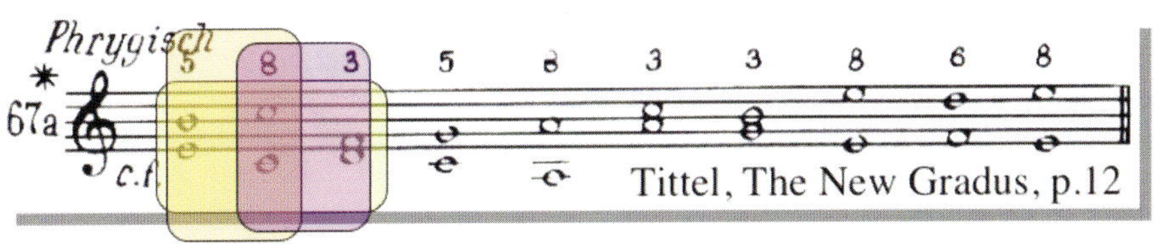

Fig. 19.10. Sequences of contrapuntal intervals must comply with the rules, such as forbidden parallels of fifths. But we also have to look at sequences of three successive intervals to grasp hidden parallels of fifths.

Figure 19.10 shows a sequence of two intervals that must comply with the rules, such as forbidden parallels of fifths. But we also have to look at sequences of three successive intervals to grasp hidden parallels of fifths. This requires a set of six notes. So we need "charts" of maximally six notes to be positioned in general position to see them all at once as a structural unit.

The other basic theory is harmony. Here, we have the most complex basic situation which is tonal modulation. Let us follow this theory in the lines exposed by Schönberg in his Harmony treatise [123], and take an example from page 197 (figure 19.11).

The example shows a covering of the modulatory phase by two charts of three triads each. The first one contains the modulation process *Cmajor* →

Fig. 19.11. A modulation from *Cmajor* to *Fmajor*, showing the relevant charts for the tonal transition, followed by a cadence.

Fmajor as described by Schönberg: neutral degree IV in *Cmajor*, pivotal chord II in *Fmajor*, tonic if *Fmajor*. And the second chart shows the cadence of the new tonality with the two chords $IV_F^7 = IV_F \cup VII_F, I_F$. So the maximal chart here has nine tones.

This means that we have an overall maximum of nine tones to be put in general position. This is, of course, a very rough requirement, since we do not represent the precise contents of these note assemblies, but only the geometric framework where to position them. But we believe that this geometric action is crucial for the best representation of theoretically important note groups.

With this in mind, we now have the following corollary of the above theorem:

Corollary 1 *We need at least four instruments of the violin group to position nine notes in general position in order to represent the essentials of contrapuntal and harmonic thinking in the late eighteenth century.*

This follows at once from the inequality $9 \leq n + 5$ from Theorem 1 of section 19.4.

Therefore, four instruments from the violin family are sufficient; of course, more can be used, but they are not necessary. *I argue that this is the deeper reason why Haydn, Schumann, and others did not need a fifth voice in these string instrumentations.*

String quartet theory therefore turns out to give a beautiful account on the question of expressing musical contents with a specific instrumental ensemble. It does so not only on a theoretical level of musical symbols, but also on the level of performance as a subtle shaping of sophisticated instrumental parameters of instruments that cannot be understood and played except by highly trained and cultivated performing musicians.

Part IV

Rubato: Model and Software

20

Performance Scores

Those who do not read are no better off than those who cannot.

We have described the Rubato concept in section 15.3. The software Rubato was first implemented in Objevtice C on NEXTSTEP from 1992 to 1996 by Oliver Zahorka and the author. Later, the software was ported in 2001 to Mac OSX by Jörg Garbers in a research project at the TU Berlin. This version is available from [118]. Standalone versions of the MeloRubette and the MetroRubette have been developed by Chantal Buteau [10] and Anja Volk-Fleischer [106], respectively.

In these implementations, rubettes were quite large components. Their in- and outputs were very limited; they were designed for well-defined purposes of performance theory. No components for composition were available, although this was planned. The project was terminated long before the full potential had been realized. Fortunately, a later research project enabled a radical redesign of Rubato. In 2006, Gérard Milmeister reimplemented the entire Rubato environment in Java (instead of Objective C), but the redesign was dramatically more flexible. This implementation can be downloaded as an open source software from [118]. It is named *Rubato Composer* in order to distinguish it from the OSX version, which we sometimes call "Classic Rubato".

Essentially, the rubettes are now more specialized and may be connected at will, since the input and output of rubettes are alway denotators, so the general data format developed for Rubato is the lingua franca of inter-rubette communication. It also has been extended in that now the mathematical basis of denotators are general points in the topos of presheaves over the category of modules. This means that we can consider arbitrary morphism on modules into contravariant functors on the category of modules as arguments of denotators. Since this is not of more specific relevance in the present state of performance theory, we omit details. But it is important to know that this version needs a portation and redesign of analytical and performance rubettes. In contrast,

G. Mazzola, *Musical Performance*, Computational Music Science,
DOI 10.1007/978-3-642-11838-8_20, © Springer-Verlag Berlin Heidelberg 2011

composition is now taken care by very powerful rubette constructions by Karim Morsy and in particular the BigBang rubette by Florian Thalmann [97, Chapters 16, 17]. Some rubettes for the recomposition of Boulez's *structures pour deux pianos* have been implemented by the author [92].

Let us now concentrate on the performance-oriented parts of Classic Rubato. We have already discussed the analytical rubettes for rhythmical, motivic, and harmonic analysis, as well as the PrimavistaRubette for score-defined default performance constructions. We shall now look at the core rubette: the PerformanceRubette and its operators.

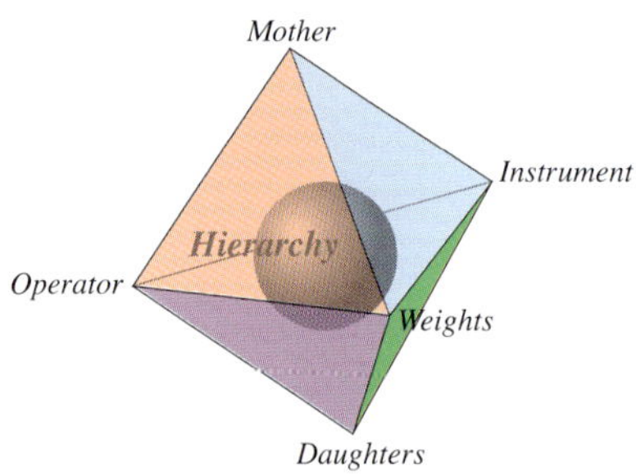

Fig. 20.1. The symbol for a local performance score (LPS).

This rubette is built upon some complex conceptual constructions created for the representation and management of full-fledged performance deployment. The first of these concepts is the completion of the information needed to define a performance. We recall that we have established the structure of a performance hierarchy. It contains the diagram of performance cells, each of which is defined by a frame, a field, a symbolic kernel, an initial set, and an initial performance.

But we have not included anything that would help construct such cells. These agents are four types:

1 On the one hand, we have to introduce the operators that shape performance, which we have discussed in chapter 17.
2 We then need the (analytical) weights that are used to make the operators act in the function of analytical data.
3 Next, we have to account for the 'mother' performance, from which the present one was deduced.
4 One additional criterion is the instrument that is used for that performance. For example, this can be a MIDI instrument.
5 It may happen that further performance hierarchies are deduced from this one by their own operators, etc. We call them the daughters.

The total information gathered with these specifications is called a *local performance score* and denoted by LPS (figure 20.1).

On Rubato, we have implemented a number of operators, which we had mostly discussed in chapter 17: splitting operator, symbolic operator, physical operator, and field operators. The field operators are (at present) the tempo operator and the scalar operator, the latter being a far-reaching generalization of the tempo operator, which we shall not discuss here since we do not use it.

This basic performance structure is dependent upon the mother. This mother can be either another LPS that has already been constructed or else primary information given by a score file. The LPS would then use the primavista operator to generate a first performance.

21

Stemma Theory

O matre pulchra filia pulchrior.
Horace (658 B.C.)

The Rubato methodology was opened towards a variation of performance rules and their rationales, and although only the analytical rationale has so far been implemented, there is no obstruction against enrichments with emotional or gestural expressivity. This is due to the modularity of the Rubato architecture. Performance rules are by no means encoded in the analytical components. This is a categorical improvement on the KTH approach, which takes, for example, the harmonic charge rule as a compact, undivisible rule that cannot separate the analytical part (in this case, it is the calculation of harmonic charge) from the shaping commands. This is reflected in the structure of shaping operators. They do certain things with whatever weight comes as input. For example, the tempo operator would take any weight and apply it to the given articulation field via the linear operator $Q_w(E, D)$ as described in chapter 17. The weight can come from rhythmical, motivic, or harmonic analysis, or even be a weighted mixture thereof. This is taken care of in the PerformanceRubette's *WeightWatcher unit* (figure 21.1).

Fig. 21.1. The WeightWatcher window of the PerformanceRubette enables the input of a weighted combination of different weights for the chosen operator.

G. Mazzola, *Musical Performance*, Computational Music Science,
DOI 10.1007/978-3-642-11838-8_21, © Springer-Verlag Berlin Heidelberg 2011

Even more importantly, the Rubato methodology is also open toward the multistep performance approach. In fact, the KTH approach is a one-step procedure: We apply a bunch of operators (rules) to a given input (the primavista performance) and then obtain the final performance. There is no internal mechanism to concatenate a number of performances. The changes defined by those rules are "stateless," so to speak, they have little if any (some of the rules might act on a previously shaped performance, but most don't) consciousness of the previous shaping results. This is, however, far from the human approach of rehearsing.

But let us first discuss the philosophy of rehearsal! Rehearsal is not only a technical process. It is above all a logical evolution and an unfolding of understanding in time, be it in the individual process of working out a performance or be it in the historical evolution of the understanding of a composition and its performances.

That is to say that rehearsal is not a sign of imperfection, but the silver bullet to the X-ray of the score's innervation as imagined by Adorno.

The logic is that performance is multilayered, since the rationales and operators regard not all parameters at once, and since certain deformations are more basic than others, e.g. the physical operators should come after the field operators. And it is hierarchical in the sense that certain deformations regard larger portions of the score, whereas others regard more local parts, like global tempi and local agogics or the general emotional "mood" of a piece that preceeds the local changes. The entire logic is also a type of syllogism: If this is done to the score, then do that next.

This insight can be formalized by an evolutionary architecture. We model performance as a process of genealogical development, similar to the biological situation. In this sense, earlier stages of evolution are not just less perfect, but also necessary in the logic of steps to more complete organisms (which for humans is sometimes not so clear...). We call the genealogical formalism of performance evolution the stemma. So we have to think about generating stemmata as an evolutionary process in performance.

This being so, the temporal stemmatic consecution is also one of evolutionary logic. The historical dimension of rehearsal is—besides the trivial technical aspect—the trace of an evolutionary process.

Rehearsal is a deployment of logical understanding in time, much as composition is the gestural deployment of the compositional formula.

The big problem here is the communication of such a historical unfolding in a single performance! How can one communicate the stemmatic construction from its resulting output? Should one—in the limit—play the entire stemma? The question is similar to the question of how to perform a Schenkerian analysis in order to make evident the Schenkerian layers in their logical implications. But our situation is more complicated since Schenkerian analysis is a simultaneity of layers, whereas the stemmatic structure is a historical process whose result is not necessarily faithful to its genesis. It is that fundamental problem in jazz: to play the tradition even in the most advanced free contcxt. And to play it not with apologies about the imperfection of an earlier stage, but with the full appreciation of stemmatically prior levels of a logical tree.

So for the time being, we understand the necessity of a stemmatic evolution in the making of a performance, but we do not fully understand how such a fact would be performable.

21.1 Tempo Hierarchies in the *presto*® Software

To illustrate the necessity of stemmatic techniques in performance, we want to discuss the implementation of tempo hierarchies in the composition software *presto*®. Tempo hierarchies are stemmata in the simple domain of tempo. They are now a model for the general stemmatic approach.

Tempo is a classical performance topic, but when modern performance researchers, such as Todd, Friberg, or Peter Desain and Henkjan Honing [26] attempted to control this phenomenon on a precise conceptual a computational level, some strange things happened. First, the very concept was put into question. Desain and Honing in [26]:"There is no abstract tempo curve in the music nor is there a mental tempo curve in the head of a performer or listener." This was perfectly in the line of the early electronic musicicians, who negated tempo altogether. Herbert Eimert and Hans Ulrich Gumpert write [27]:"Electronic music neither knows tempo nor metronome marks, but it documents its connections to the phenomenon of time by the most precise time indications which exist in music." The negation of tempo came from the confusion of symbolic and physical reality in music. But it also came from the fact that traditional performance science in music was all but adequate to performance on an artistic level. We have found in a recent book [21] a reference to the classical work "Pianoforte Schule" by Carl Czerny [17] as exclusive reference to precise tempo shaping. Let us look at that example (figure 21.2).

It shows a short exercise by Czerny, where he compares tempi according to four shapings: 1. in tempo, 2. in tempo/ un poco ritenuto/ smorzando, 3. in tempo/ poco accelerando/ rallentando, 4. in tempo, molto ritardando/ perdendo. My coworker Zahorka has reconstructed these situations with a quantitative tempo curve that can be defined in the module AgoLogic of the composition software *presto*®. The results are all unsatisfactory; hear the first three items of example ♫ 14. The shaped score sounds dull and is not what one would expect from an expressive performance.

This lack of a valid tempo theory does however not mean that there is no tempo, it simply means that one tempo curve is not sufficient. With AgoLogic, one can define very interesting tempo curves, for example *accelerando* curves for drum rolls. It works, but it fails when it comes to artistc tempo shaping. One of the evident situations where a tempo curve is insufficient is Chopin's rubato. This means that one typically plays the left hand with a certain tempo (not necessarily constant), and the right hand plays in a varied tempo, however

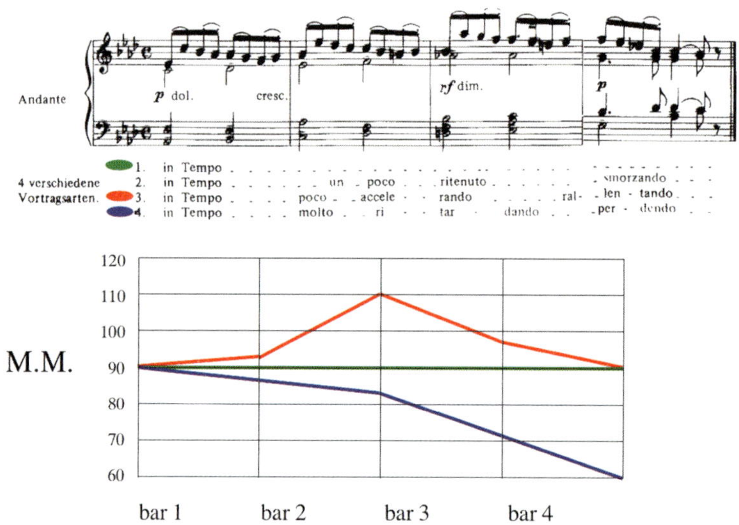

Fig. 21.2. Czerny's experiment with four varied tempi.

synchronous with the left hand tempo in all important onsets, such as bar lines. We have tried this idea in AgoLogic, see figure 21.3.

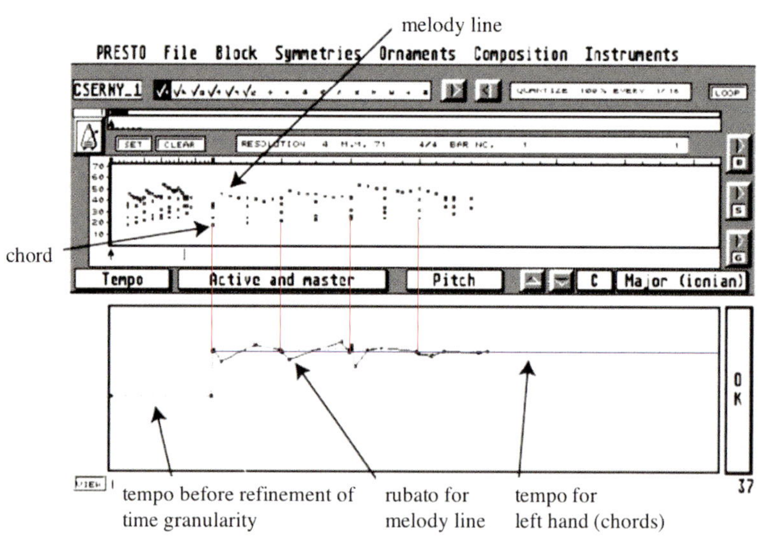

Fig. 21.3. The Chopin rubato, as represented in *presto*®'s AgoLogic module. The constant tempo of the left hand is locally varied for the right hand, synchrony being set for all bar lines.

The left hand is given a fixed constant tempo T_L. AgoLogic enables local variation of this tempo. To this end, we define a splitting of the entire time into four one-measure units. Within each such measure unit, the right hand notes are subjected to locally varied tempi. This is achieved by graphically defining one new curve per measure. The user can draw polygonal tempo curves $T_m(E)$ in each measure m; however, the software will always correct the user's curve to get an overall duration equal to that of the left hand's tempo, i.e., if D is the symbolic duration of a measure, we must have

$$\frac{D}{T_L} = \int_{E_0}^{E_0+D} \frac{dE}{T_L} = \int_{E_0}^{E_0+D} \frac{dE}{T_m}.$$

The software applies an algorithm to approximate the precise value by varying the polygonal shape drawn by the user until a limit of unprecision is reached. The result of this user-defined graphical shaping of local tempi turns out to be quite satisfactory. See ♫ 14 for an audio file (with the three versions 1,3,4 from Czerny and last our Chopin rubato version) recorded from a Yamaha MIDI grand.

The general idea behind this experiment and AgoLogic's method is to introduce hierarchies of tempi (figure 21.4). Which means that we are given "mother tempi," which have "daughter tempi." The daughters are those local variations of the mother tempi, subjected to synchrony on determined boundary points. AgoLogic enables arbitrary deep and arbitrary broad hierarchical ramifications: A mother tempo curve might have as many daughters as required, and the genealogical tree from mother to daughter to granddaughter, etc. might be as long as desired.

Figure 21.4 shows a mamma curve, which has two daughters: curva1 and curva2, and each of them has one daughter: curvetta3 and curvetta4. Each daughter has a curve whose duration is the same as that of the mother within the daughter's time interval. Since every curve in such a hierarchy of tempi is associated with its own set of notes, the entire piece has no overall tempo curve. But the hierarchy expresses a genealogy of tempo logic (this is why we coined this module AgoLogic). The daughter's tempo is a derivative of the mother's tempo, and the daughter's proper shaping is due to the logic applied to this daughter's agogical performance rationales. An impressive application of this module has been realized for a small portion of Frédéric Chopin's *Impromptu* op.29 (figure 21.5).

This short part in Chopin's *Impromptu* is a delicate performance task since it has a complex hierarchy of time layers. The basic mother tempo is the one defined by the three half notes. The mother then has six daughters, each half-note duration having two daughters, one for the left hand, one for the trill in the right hand. Each left hand's daughter has its own daughter defined by the arpeggio. This tempo hierarchy enables very different tempo configurations, from the beginner's rendition to the virtuoso, such as Barenboim or Pollini style—see ♫ 19 for an audio file with these four version: primavista, beginner, Barenboim, and Pollini.

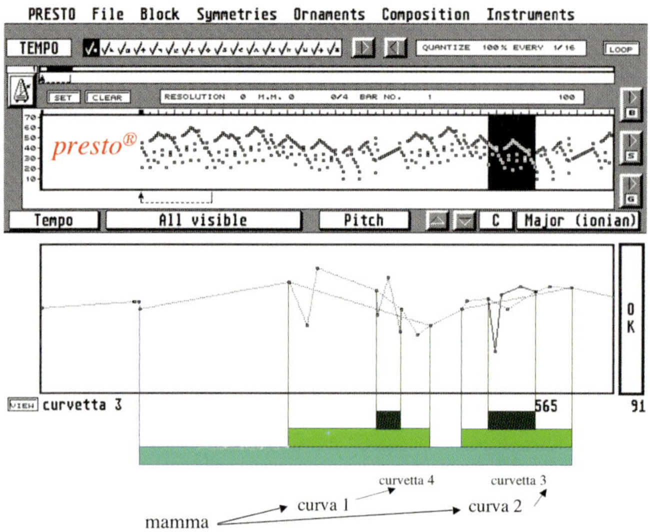

Fig. 21.4. The AgoLogic module in *presto*® models tempo hierarchies by a matrilinear paradigm of mother-daughter inheritance. A mother can have several daughters, and the genealogical tree can be of arbitrary depth.

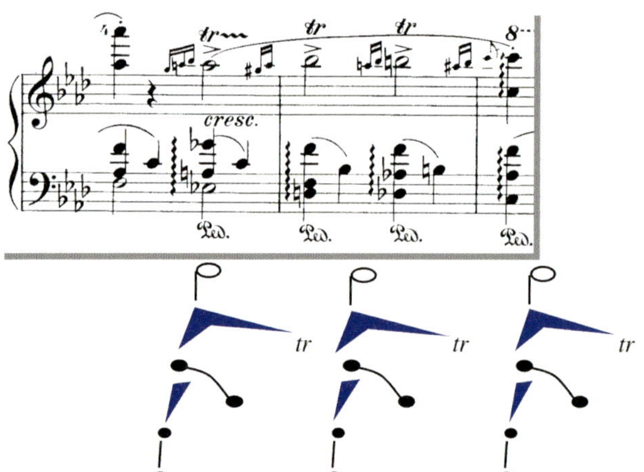

Fig. 21.5. This short part in Chopin's *Impromptu* op.29 is a delicate performance task since it has a complex hierarchy of time layers.

21.2 The General Stemma Concept

With tempo hierarchies in mind, we can now formally describe the stemmatic unfolding of a performance. It is a tree whose nodes are local performance scores (LPS) as defined earlier in chapter 20. Each non-top node is attached to its mother LPS and ramifies to its daughter LPSs. The top LPS, the *primary*

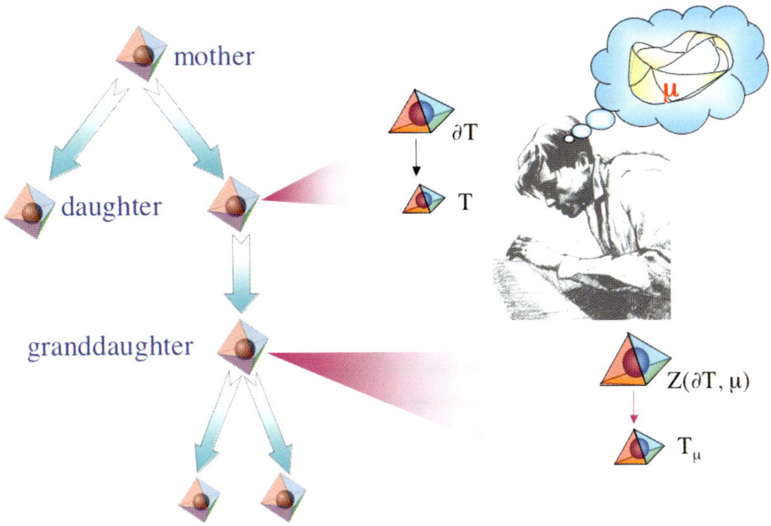

Fig. 21.6. The general stemma concept shows a tree of local performance scores. Their unfolding corresponds to the successive refinement of performance according to selected rationales. In our image, this would be Glenn Gould's harmonic thought that refines an articulation field.

mother, is attached to the score, which is not an LPS. Figure 21.6 illustrates a stemma scheme. It starts with the primary mother and ramifies to her two daughters, one of them having her own daughter (granddaughter), who in turn has two daughters. Therefore, there are three leaves of this tree, and it is those that will be played as a performance output. The others are only part of the internal processing. The passage from a mother to her daughter, as shown in this figure for the right daughter and the granddaughter, is caused by quasi-sexual propagation in that the granddaughter is generated from her mother using the granddaughter's operator, which plays the role of a father.

In our example is shown an articulation hierarchy with a parallel hierarchy $\partial T \to T$, which is reshaped by the analytical weight stemming from a harmonic analysis symbolized by μ, the harmonic Möbius strip of the triadic covering of a scale (see [84, section 13.4.2] for this analytical structure) and yields the granddaughter's articulation hierarchy $Z(\partial, \mu) \to T_\mu$. Since the stemma is a tree, there are no closed paths and no ambiguous states for multiple mothers, for example. This seems reasonable, but it prevents us from formally thinking of feedback cycles. This is an important requirement for realistic modeling of performance genealogy. The only thing we can do at present is to take the output, analyze it 'offline' and redefine parts of the existing stemma to get a better final performance. This is similar to the analysis-by-synthesis approach of the KTH school. But it differs insofar as it is not new rules that must be reprogrammed, because the very flexibility of the system enables us to choose

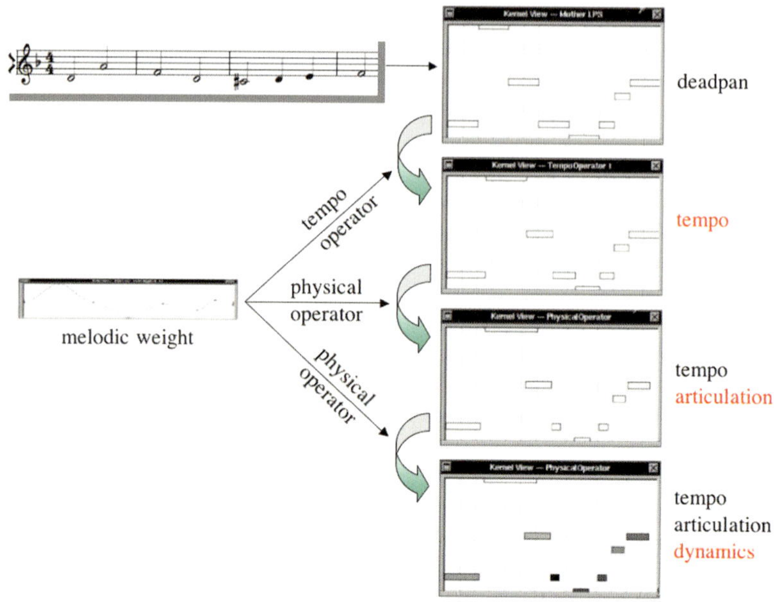

Fig. 21.7. The main melody of Bach's *Kunst der Fuge* is performed in a stemma starting with the deadpan primary mother, then generating its daughter by a tempo operator using the inversed melodic weight, then by a physical operator using the non-inversed melodic weight and acting on duration (for articulation), then by a physical operator using the inversed melodic weight and acting on loudness for dynamical shaping. The graphical representation of the score's notes in Rubato's windows here is in rectangles with horizontal position equal to onset, vertical position is pitch, width is duration, and gray level is loudess.

different parameters or weights or operators when stepping to an improved performance.

The important point in this reshaping process is that we can redefine any of the nodes' parameters at any time. So the performance history can be rewritten in any of its nodes. And, if necessary, one can also destroy a sub-tree starting from any node and reconstruct the tree below that node.

Figure 21.7 is a very simple example of a stemma, which we did on the first NEXTSTEP implementation, showing the NEXTSTEP setup for historical reasons. We take the input as being the main melody of Bach's *Kunst der Fuge*. It is performed in a stemma starting with the deadpan (no shaping, just the mechanical rendering) primary mother, then generating its daughter by a tempo operator using the inversed melodic weight, then by a physical operator using the non-inversed melodic weight and acting on duration (for articulation), then by a physical operator using the inversed melodic weight and acting on loudness for dynamical shaping. The corresponding sound example is ♫ 20.

21.3 The New Performance Rubette: Realtime, Openness, and Gesturality

The increasing potency of current computer systems facilitates a new way of representing and manipulating structural as well as analytical data. This has been exploited for music composition in recent developments for the *Rubato Composer* software. The module *BigBang Rubette*, for instance, provides a gestural interface with which results of affine and other transformations of music are visualized and sonified during the process of manipulation. This gives composers the impression of transformations being a gradual and gestural process, and it makes them accessible even for non-mathematicians. Furthermore, it reduces the ubiquitous distance between production and product.

These paradigms bear a great potential for the further development of tools for performance analysis and modeling, as well. The *MetroRubette* (chapters 16.1, 22.1), currently reimplemented by the author, uses gestural concepts for the input of its parameters, such as *local meter length* and *profile*. The resulting weights are immediately calculated whenever the parameter values are changed. These weights, output in denotator format, can then be used as an input for other rubettes for example, for those who use them to shape a performance of the piece the metrical analysis was made for.

As described in chapter 20, the architecture of Rubato Composer enables the design of smaller computational units than previous software products used. One traditional rubette is now split up into several, all dealing with a well-chosen, specific, and atomic task. This is mostly facilitated by the universal and flexible data format of denotators. Because of their generality, rubettes can be designed such that they can process inputs of any imaginable form. In this way, a rubette created for processing musical denotators in a general manner might as well be used for image processing, as soon as appropriate forms are defined. As described in [135], this type of open design, paired with the generality of the denotator format, ensures maximal flexibility and modularity. The more universal their elements are defined, the more modular the potential networks will be. Such rubettes can be seen as analogous to the musical agents described in section 4.11. They are "performers" themselves that receive a score (neutral niveau), process it in their own way, and pass on a new score that reflects their operations. To reach the highest possible expression, it is important to keep the ears as widely open as possible. Only true musical hearing will define a collaborative space of maximal extent [91].

The next step in the development of Rubato Composer will be to enhance communication between rubettes using concepts from gesture theory. For this, we will need to redesign the Rubato Composer communication system, so that it not only handles offline messages triggered by the *run network* button but also streams of messages, the rate of which would be defined by the outputting rubettes themselves. This way, if a parent rubette changes its input values or its properties, all dependent rubettes will receive the newly calculated output immediately and in turn recalculate their results. This improvement will lead to entirely gestural networks of rubettes, which all react to one another in an immediate way. For instance, a rubette that calculates a performance from specific weights will recalculate the performance and ideally play it back immediately, in a similar way as in the BigBang rubette.

The two requirements formulated in this section—the openness and generality, as well as the gesturality of the communicated data in Rubato Composer—will lead to a comprehensive and flexible system for music composition, analysis, and performance. Just as openness and gesturality form the foundation of human conceptualization and experience.

An example for the application of these paradigms would be a new implementation for the calculation and manipulation of performance fields, which were introduced in chapters 8-10. A future software project aims at a tool for visualization of such performance fields. It is inspired by the *EspressoRubette* (chapter 10) and works in a similar way but is enhanced by several realtime and gestural characteristics. It will visualize the quality of a musical performance during the performance itself. If the input reference composition is an unshaped original score, the performer will see how her interpretation deviates from the original in tempo, pitch, dynamics, and articulation, each of which will be represented by an individual color field, which could be projected on a big screen. This setup can be used as a practical reference for music education, for instance, where students and teachers will be able to see immediately and precisely how their performance is shaped. The visualized performance field could then be modified with gestural interaction and played back for investigative purposes. This way, high-quality performances could be produced in a combined procedure of playing and gestural shaping and then be either saved or played back. Again, this module could then communicate in realtime with other rubettes, which could produce detailed analyses of the resulting performance or use it for music composition.

22

Case Studies

Learning by Doing.

22.1 Schumann's *Träumerei*: The First Performance Experiment with RUBATO®

Our first longer performance was constructed in 1995 with Robert Schumann's famous *Träumerei*, the seventh *Kinderszene* in his collection op.15. It was an experiment conducted in the context of a performance conference at the KTH, where different approaches to performance were compared [82], and from where we take the following presentation. The performance was played on a MIDI Boesendorfer Imperial grand piano at the School of Music in Karlsruhe. This piece was chosen because we have the detailed analysis of agogics as measured by Bruno Repp from 28 famous performances [111] by, among others, Marta Argerich, Vladimir Horowitz, and Alfred Cortot.

For this experiment, we made a rhythmical analysis by the MetroRubette and a motivic analysis by the MeloRubette. The HarmoRubette was not implemented in those days.

The rhythmical analysis is shown in figure 22.1. The parameters for these weights are minimal admitted local meter lengths = 2 and profile = 2. We see from top to bottom the weights $metroWeight_{LH}, metroWeight_{RH}$, and $metroWeight_{BH}$ for the left hand, the right hand, and for both hands, respectively. We recognize the markedly different profiles of these three weights, a phenomenon already observed in our previous discussion of the composition's rhythmical analysis in section 16.1.

For our performance, we have these weighted combinations of metrical weights for the left- and right-hand shaping:

G. Mazzola, *Musical Performance*, Computational Music Science,
DOI 10.1007/978-3-642-11838-8_22, © Springer-Verlag Berlin Heidelberg 2011

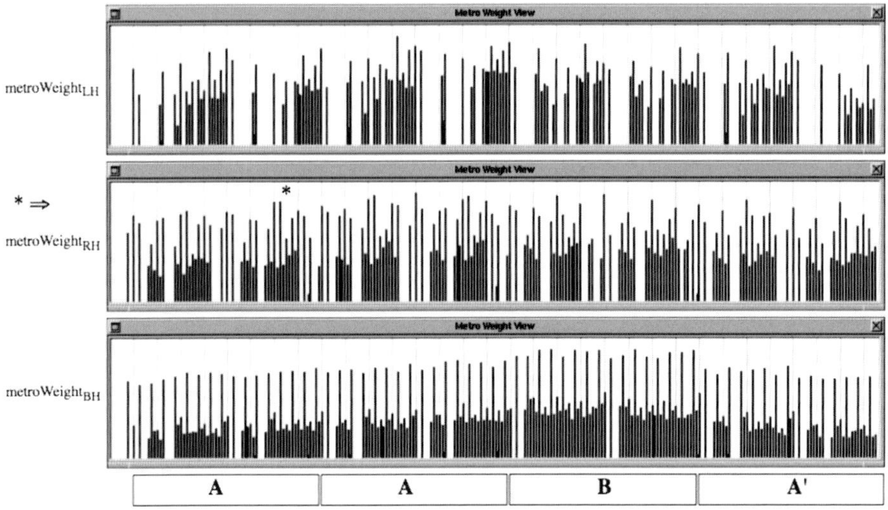

Fig. 22.1. The metrical weights $metroWeight_{LH}, metroWeight_{RH}, metroWeight_{BH}$ for the left hand, the right hand, and for both hands, respectively, with minimal local meter length = 2 and profile = 2 for Schumann's *Träumerei*. The star marks the harmonically important point where the secondary dominant appears.

$$metroComWeight_{LH} = 100\% metroWeight_{LH} + 10\% metroWeight_{BH}, \; and$$
$$metroComWeight_{RH} = 100\% metroWeight_{RH} + 10\% metroWeight_{BH}.$$

which is a strong account on the original-handed contributions, plus a small account on the combined rhythmical structure. As already discussed in chapter 16, these discrete weights are always interpolated by cubic splines. In figure 22.2, we see the splines for the weights

$$metroComWeight_{LH}, metroComWeight_{RH}$$

and their difference. We have also chosen the high and low limits of these spline weights to be 1.2 and 0.9, respectively.

The melodic analysis used here was done using the so-called *elastic* paradigm of motivic similarity. This one looks at the slopes of the lines connecting successive notes and the relative Euclidean lengths of these connections, so it is a very geometrical paradigm. Comparison among motives would use similarity of that elastic data. The similarity limit ϵ discussed in section 16.2 was chosen to be $\epsilon = 0.2$. We also decided not to compare inversion or retrograde or retrograde-inversion of such motives, but only the given motives. We also chose a small window of motives, namely only motives of two, three, and four notes each and having their note onsets between one-half measure length. This yields 1574 two-note, 1465 three-note, and 71 four-note motives. At that time, the selection of significantly more motives would have exceeded the calculation power of a NeXT computer. The graphical representaton of

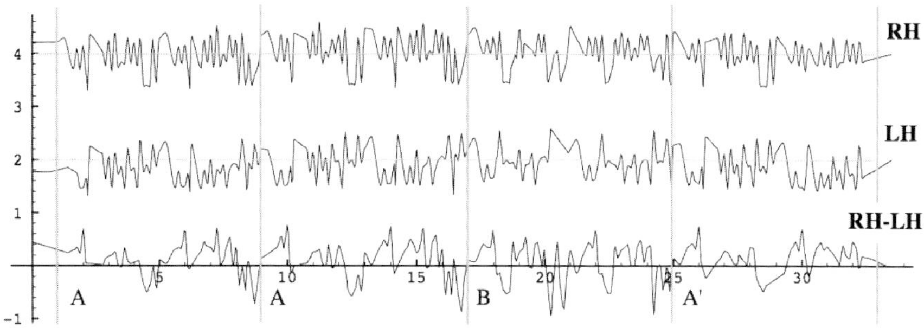

Fig. 22.2. Splines for the weights $metroComWeight_{LH}, metroComWeight_{RH}$ and their difference.

this weight is shown in figure 22.3. Since we also applied the melodic weight to shaping agogics, we needed a boiled-down version of the melodic weight, which is a function of onset only. This function just adds all note weights of notes with given onset.

The melodic weight and the inverted (!) spline of the boiled-down motivic weight is shown in figure 22.3. We have chosen the *inverted* spline between high and low limits, 1.2 and 0.9, because for some operators it was reasonable to have low influence for high melodic weights. For example, agogics should go down for high motivic weight. It is interesting to see the performance field

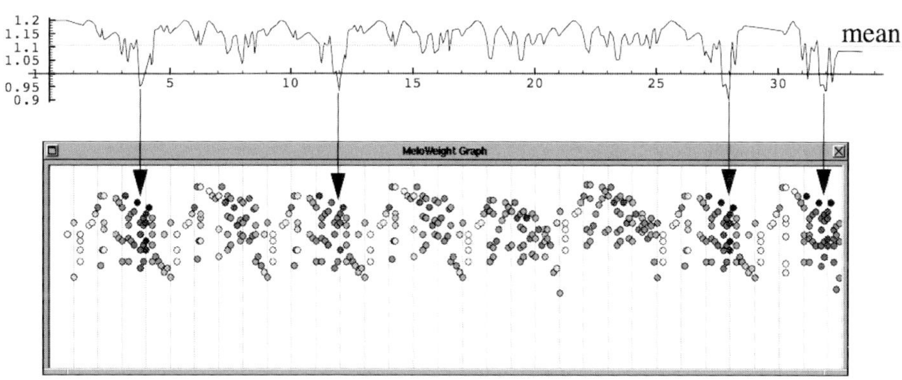

Fig. 22.3. The motivic weight $meloWeight$ for both hands (bottom) shows a markedly high weight at the end of the piece, in the repeated ascending motif $g - a - b_\flat - d$, as an important melodic instance. Above, we see the inverted boiled-down motivic weight spline.

of articulation being constructed by the tempo operator, as is shown in figure 22.4.

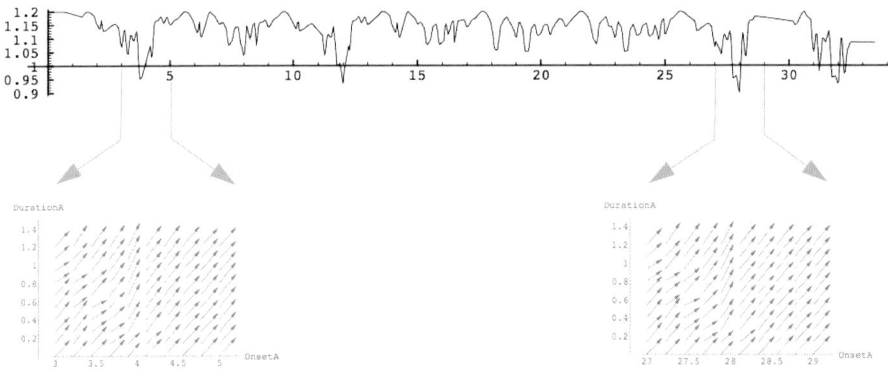

Fig. 22.4. The articulation fields generated by the tempo operator by use of the inverted bolied-down melodic weight.

Besides these analytical weights, we have created also primavista weights for agogics, dynamics right hand, and dynamics left hand. This is shown in figure 22.5.

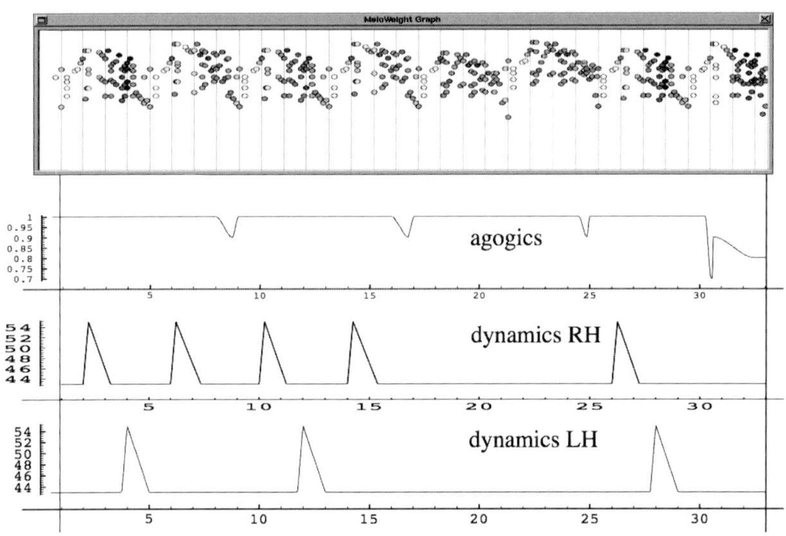

Fig. 22.5. The three primavista weights.

Using this data, we have then constructed a simple stemma as shown in figure 22.6. It splits left from right hand, then applies primavista shaping, then the physical operator (called brute operator at that time) to dynamics—using the splined weights from

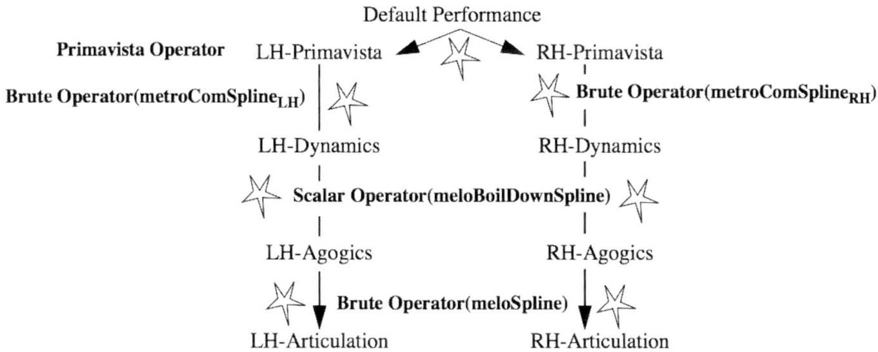

Fig. 22.6. The stemma of our performance of *Träumerei*.

$$metroComWeight_{LH}, metroComWeight_{RH}$$

then the tempo operator (called scalar operator at that time) with inverted boiled down-melodic weight, then again the physical operator for articulation—this time using the melodic weight, not the inverted boiled-down version, to the single notes' durations.

What is the relationship between the described RUBATO® performance and the well-known performances by famous artists? This question turns out to have an remarkable answer. To deal with the empirical data, we refer to the Repp's paper [111]. Repp has measured the tempo curves by the measurement of the IOI (intertone onset intervals, which is a discrete measure for 1/tempo). He than applied a statistical factor analysis to the first eight measures of the 28 performances and got four significant factors. Three of these factors turned out to be represented by high loading for a group of artists. The first factor is that shared by a large number of artists, among them Alfred Brendel. The second factor is led by Horowitz, and Repp therefore calls it the "Horowitz factor." The third factor is called the "Cortot factor" for analogous reasons.

We have analyzed the three timing patterns corresponding to the three important factors. It turns out that the situation for the Horowitz factor is in remarkable coincidence with the agogics obtained by the RUBATO® calculation, i.e. by the agogics deformation via the tempo operator from the boiled-down melodic weight. We are comparing the tempo curve of the Horowitz situation as it reads when the discrete data are completed to a cubic spline (the same method was used for the melodic weight spline), see figure 22.7. The upper curve is that of the RUBATO® agogics, then comes that of a prototypical Horowitz timing, and the third one is the product of the derivatives of the two agogics. It is negative if the slopes of the two candidates are contrarious. This shows that the fitting quality of the two curves is extremely good. This fact can also be seen by visual inspection. To say more about the rare discrepancies, we would have to go back to these recordings on one side, and do harmonic analysis on the other. But it is clear that the agogics of the

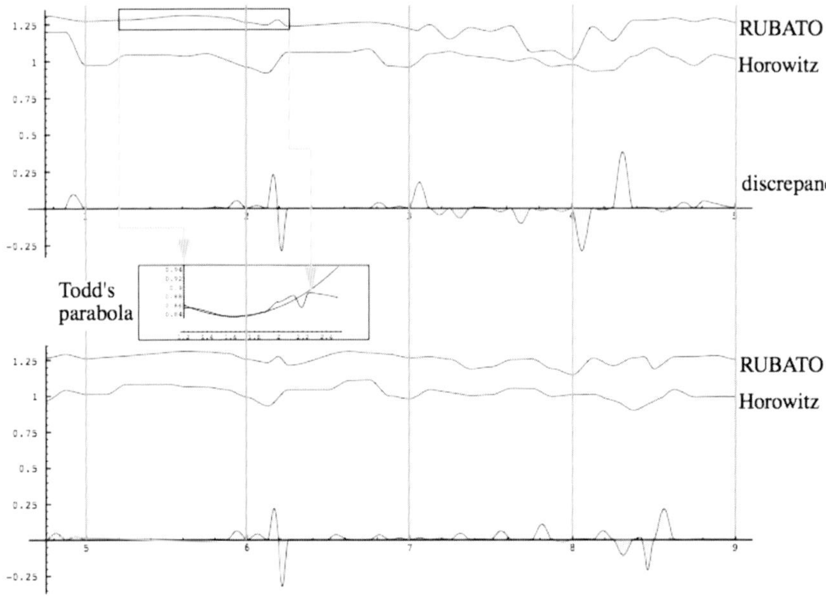

Fig. 22.7. Comparison of the RUBATO® tempo curve and the Horowitz curve measured by Repp.

RUBATO® performance—though it has a very simple stemma—is in the line of one of the most profiled styles as described by Repp.

Let us also notice that the parabolic accelerando in measures 1 to 2 discussed by Repp in the light of Todd's hypothesis [137] is not in contradiction with the RUBATO® agogics in this location. From a mathematical point of view, it is by no means clear that a parabolic accelerando is the only reasonable solution. This issue has to be settled in the light of a systematic inverse performance theory, but see chapter 24.

22.2 Schumann's *Kuriose Geschichte*: The First Analytically Complete Performance Experiment with RUBATO®

A still more realistic second example of a stemma is shown in figure 22.8, the very first *extensive* experiment in Rubato-driven performance we did in 1996 on the MIDI Boesendorfer Imperial grand at the School of Music in Karlsruhe. The deadpan version (without any performative shaping of the score data) can be heard in example ♫ 21; the final performance is documented in ♫ 22.

Let us look at this historical example of a stemma: the stemma for the composition *Kuriose Geschichte*, the second *Kinderszene* in Robert Schumann's synonymous collection op.15. This stemma was constructed for the

NEXTSTEP RUBATO® by the author, Oliver Zahorka, and Joachim Stange-Elbe. It took us three days to realize the whole setup and performance. The performance of the piece is documented on ♫ 22, and in a broadcast of the Austrian TV [34]. Although the stemma is quite primitive, the shaping results were satisfactory and taught us a lot about the empirical aspects of computer-assisted performance research. In particular, we learned that it can be very difficult for humans to listen dozens of times to successive and only slightly altered versions of a performance. At the end of a day of such work, one cannot tell anymore what matters and what is really different or just imagination, even for three independent listeners!

Although each single refinement layer is controlled by one and the same operator (horizontal arrow), each daughter had to be performed as an isolated instance, since no grouping methods were implemented. The construction of this stemma first follows the splitting of right (RH) and left hands (LH), then, after the shaping of primavista dynamics and agogics, global agogics is constructed on these two LH and RH symbolic kernels. The splitting for operators $\Omega_5, \Omega_6, \Omega_7$ regards a small number of measures that have to undergo a more differentiated rubato. The final shaping regards fine "tuning" of dynamics and articulation in all leaves.

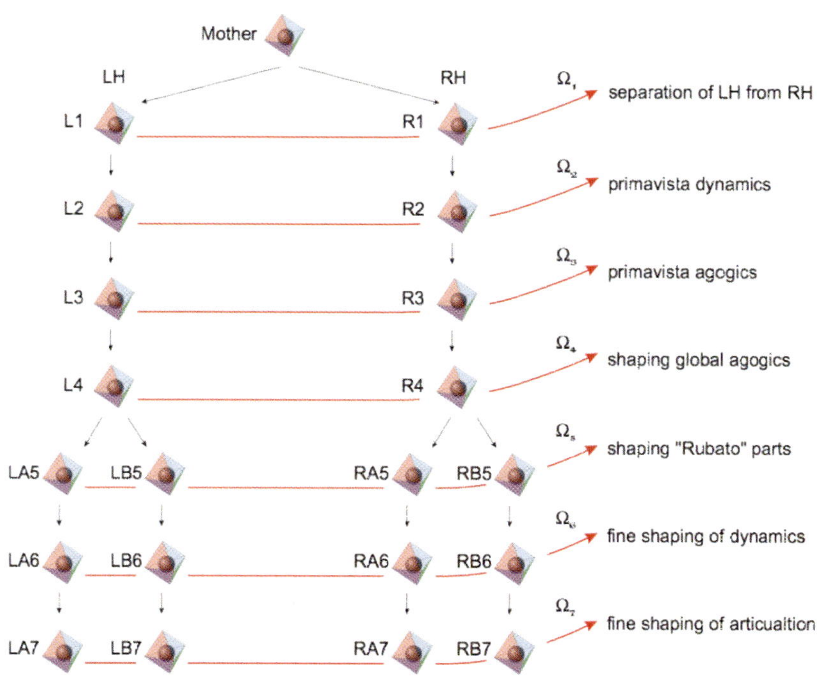

Fig. 22.8. The stemma of the first RUBATO®-driven performance construction of Schumann's *Kuriose Geschichte* in 1996.

22.3 Joachim Stange Elbe's Performance of Johann Sebastian Bach's *Kunst der Fuge*

Fig. 22.9. First page of Bach's *Kunst der Fuge*.

Joachim Stange-Elbe has investigated Bach's *Kunst der Fuge* (figure 22.9) in detail and created very convincing performances thereof with RUBATO®. The deadpan version of its *contrapunctus III* is documented in example ♫ 23, while the final performance can be heard on example ♫ 24. We however do not include this work as a further example of performance construction in a technical sense, rather we focus on the problem of analytical performance as such. What is the relation of analytial weights, operators, and the aesthetics of performance?

So, let us first look at the premises used by Stange-Elbe.

This experiment is fully accounted in [130], here, we give a concise presentation. The version of RUBATO® used in this experiment is the one compiled for OPENSTEP/Intel. The *contrapunctus III* in Bach's *Kunst der Fuge* has these characteristics: It is a four-voice composition, comprises 72 measures, has time signature 4/4, and has tonality *D*-minor. The main theme of *Kunst der Fuge* is only used in its inversion and appears the first time in a rhythmically dotted and syncopated variant; the fugue starts with the theme in its comes shape and contains three complete developments (measures 1-19, 23-47, and 51-67).

We give an account of the rhythmical and melodic analyses, whereas the harmonic analysis has not been done. Stange-Elbe decided that Hugo Riemann's theory, which is implemented in the HarmoRubette, is not suited for Bach's harmonies. He argues that when using the Riemann theory, which was developed from the Viennnes classics, the specific harmonic structures of a contrapuntal maze, where harmony does not result from progression of fundamental chords but from the linearly composed voices, can be captured only in an incomplete way.

22.3.1 Rhythmical Analysis

For the rhythmical analysis of the *contrapunctus III*, the calculations were made for each single voice, including the sum of the voice weights, and for the union of all voices. The settings of the weight parameters are these: metrical profile is 2, quantization is 1/16, distributor value (the weight factor for weighted sums of weights) is 1. Since the metrical profile of all voices should be viewed under the same valuation, the distributor value was set to a common neutral value; the value 2 for the metrical profile resulted from several trials of analyses and yields a balanced distribution of the weight profile.

The value for minimal length of local meters was successively decremented starting from the length of the largest local meter and descending until value 2, where the smallest cells are caught in their signification for the metrical overall image.

22.3.2 Motif Analysis

For the calculation of motivic weights, each single voice of the *contrapunctus III* was analyzed separately. Stange-Elbe refrained from a motivic analysis of the union of all voices because by the contrapuntal structure of the single and autonomous voices within the polyphonic setting, a motivic setup across the voices seemed rather unlikely and therefore was omitted.

The settings for the motivic analysis were chosen as follows: Symmetry group: counterpoint, which means that motives also were compared to inversion, retrograde, and retrograde-inversion of other motives; the similarity paradigm was chosen equal to that in our discussion of Schumann's *Träumerei*, namely elastic; the tolerance number was set to $\epsilon = 0.2$. By the choice of the counterpoint symmetry group, the theme forms *recta* and *inversa*, as well as their (possibly appearing) retrogrades, were considered as being of equal weight. The neighborhood value has been chosen as based upon analytical experiments during the development period of RUBATO®.

As to the values for motif limits, compromises with the calculation power had to be made. By making the span[1] equal to 0.625 and setting the cardinalities of motives from 2 to 7, motives within a span of a half note plus a quaver were captured; this corresponds exactly to the duration of the theme where the transition of the virtual theme to the interludes must be recognized. With the results of the metrical analysis, some regularities in the microstructures can be read at first sight; herein we find in particular the onsets of the theme within a particular development.

While further considering these weights, the overly long pauses in the soprano, tenor, and bass voices attract attention. Further, in the length proportion of the single weight representations, the succession of onsets of the single voices (tenor-alto-soprano-bass) is reflected. Moreover, a significantly lower motivic profile at the beginning and after the longer pauses of the respective weights can be observed—due to preceding pauses, this is the case of exposed thematic onsets.

For the weight values, a neat exposition of the inverted gestalt of the original theme is observed, bearing nearly identical weights at the beginning of every motivic weight. Here even the differences of comes and dux forms are visible, since the weights of the tenor (first appearance) and soprano (third appearance) differ slightly by the different initial interval of the theme (descending fourth in the comes and descending fifth in the dux form) from the weights of the alto (second appearance) and bass (fourth appearance).

Other clearly visible onsets of the theme in inverted shape are recognized after the long pauses in the soprano (eighth appearance), bass (ninth appearance), and tenor (twelfth appearance). Characteristically, the inverted shape always appears after pauses.

At first sight, these observations may seem to be tautological. However, if these weights are viewed with respect to their sense and purpose and their force to shape performance, then the transition from a quantitative to a qualitative information content becomes evident. Thus the different onsets of themes can be shaped by these weights in one and the same way; if these weights are used—in inverted form—for the dynamic shaping, then the thematic onsets can be stressed with plasticity.

[1] This is the maximal admitted distance between first and last onset of a motif.

22.3.3 Target-driven vs. Experimental Stemma Constructions

Before the stemmatic construction for *contrapunctus III* is discussed, some general remarks regarding the various performance strategies are necessary. In the course of the single performance parcours, two different approaches resulted that would turn the given analytical weights into expressivity: the *target-driven* and the *experimental* strategies.

The target-driven strategy has its roots in the knowledge about existing performances; it is stamped by a preliminary experience of how the piece should sound and has been performed. With this procedure, the weights are used in a way that targets a predefined performance. One—just to name a pithy example—was oriented towards Glenn Gould's Bach interpretation; the corresponding weights were selected according to these targets to obtain particular effects. In this procedure, however, the intrinsic structural meaning of analytical weights was ignored! Stamped by the knowledge and the expectation of the existing performances, this strategy did not allow one to judge and categorize those performance constructions that did not suffice for the music-esthetic exigencies.

The other approach, the experimental strategy, moves the analytical weight to the center in order to investigate how this weight could 'sound', and which analytical insight it could convey in the listening. With this procedure, which views the main performing agent entirely within the weight, one has to free oneself completely from horizons of expectation for any particular performance target. The working process on such performances, the acquaintance of experience with the most different weights, and the playing with their effects taught us in the course of many experiments that this strategy would give rise to much more interesting performance aspects. Here we also have the freedom to admit extremal positions that disclose more about the inherent musical structure and as 'daring ingredients' may evoke lively musical expression.

Moreover, the experimental approach to single performance aspects, which starts from curiosity about the sonic realization of analytical weights, conveys a deeper insight into to score's musical structure. This path has its take-off in a "sonic analysis," or else in "the sonic analytical structure" and aims at a "musically reasonable performance." It is centered around the researcher's curiosity for a sounding and interpretational realization of analytical weights and for "the never heard," and it is paralleled by a liberation from expectational presets. Moreover, this strategy tries to apply as few weights as possible in order to couple the clearest possible analytical statements with the resulting performance.

22.3.4 Performance Setup

The performance of *contrapunctus III* took place in three parcours. The first one was entirely devoted to a target-driven strategy centered around the shaping with a single weight in order to sound the potential of a single weight. The global application of weights and the usage of a single weight showed its limits. For example, the global application of weights failed in the different grades between the contributions of the four voices. Especially with the motivic weights of the tenor and bass voices, different weight profiles become visible that cannot be eliminated even by suitable deformations. These differing profiles of weights result from the compositional structure. As this one splits into a number of parts—developments and interludes, groupings by harmonic closes and semi-closes—the division of the voices according to such compositional criteria is legitimized. Within these parts, the selected weights can be applied with different intensities and thus equalize the disparate shapings.

The subsequent parcours switched to an experimental strategy, which yielded much more successful and conclusive results. Nonetheless, all these approaches contributed results that influenced the final result in a significant way.

Generally speaking, the procedure in all these parcours first focused on isolated single aspects of performance (articulation, dynamics, agogics) and then were put together for the final parcours. For the complete description of all these steps, see [130].

22.3.5 Construction of Third Performance Parcours

Because of these different dynamical profiles, the principle of former performance experiments—the exclusive usage of a weight and its global extension—had to be given up. In a first step, it was recommended to split the single voices at appropriate locations, and in a second step, a regress to the metrical weights already used in the first parcours and their renewed application under other viewpoints (a mixed usage together with motivic weights) seemed reasonable. The shaping of articulation from the second parcours would be conserved.

In a preliminary step, a division of the single voices had to be executed. To this end, one had to find structurally legitimate points from the musical context, such as articulation by harmonic incisions or thematic groupings for developments and interludes.

The first division of all four voices took place in measure 39, legitimated by a harmonic close to the major parallel of the minor dominant (C-major); at the same time this is viewed as a possible ending of the second (however incomplete) development and a beginning of a four-measure interlude.

In order to equalize the dynamical unbalances relating to the interludes from measure 19 and 46, a further division of the two halves of the fugue was necessary. A division of the first half was recommended in measure 19, having a close of the first development (exposition of fugue) and its half close on the dominant (A-major).

Because of the too-strong dynamic sink of the three-voice interlude from measure 46/47, the division of the second half had to take place no later than at this point. This division was legitimized by the half close on the minor dominant (A-minor) beginning in measure 46 on the one hand, and the simultaneous ending of the second (then complete) development according to the three-part construction of the fugue.

For the subsequent performance shaping, consider figure 22.10. Besides the already known preparatory steps—horizontal division into single voices (Level 3) and equalizing of loudness (Level 4)—two performance steps for the later shaping of global agogics were inserted (Levels 5 and 6). This trick is applied because agogics needs long calculation time on the global level of single voices and should be calculated after the stemmatically subsequent shaping articulation and dynamics. The vertical division of the single voices is applied in the previously described steps (Level 7 and 8). For the subsequent shaping of articulation and dynamics, each voice had to receive its separate and individual performance shaping for the four sections. This enabled us to apply different parameter values for the intensity effects, one per used weight.

For the shaping of articulation, the three already elaborated performance steps were inherited.

As is seen in the stemma (figure 22.10), the shaping of dynamics was realized in three consecutive steps. Here, besides the known motivic weights, two additional metrical weights were applied. For the first step (Level 10), we applied the metrical weight from the union of all voices with minimal length of local meters equal to 2, in inverted form, and without deformation.

Upon this stemma, the second step (Level 11) applied the metrical weights with value 5 for minimal length of local meters for each individual voice in inverted form and also without deformation. For the concluding shaping of dynamics, the already known motivic weights were applied to give the thematic onsets a plastic relief.

The result of this performance communicates a relatively balanced dynamics, spread over the whole contrapunctus; the thematic onsets gain a profile, which can also be confirmed in the slight crescendo that leads to the beginning of the third development after the three-voiced interlude (from measure 46/47).

Bringing together the dynamic and the already elaborated articulatory aspects, the result can be stated as a complementary shaping of both performance aspects, which on top of that reveals a musical sense in the elaboration of thematic onsets and the three-voiced passages of the interludes.

For the shaping of agogics, the said levels 5 and 6 of our stemma were reserved. Stange-Elbe did two different subsequent performance parcours with two different metrical weights: the sum of all voice weights (minimal length of local meters: 2) and the weight of the voice union (minimal length of local meters: 91 (!)).

22.3.6 Final Discussion

In the course of the performance experiments, two different approaches and performance strategies crystallized. Stange-Elbe tried to give the score's text an immanent shaping by means of two approaches:

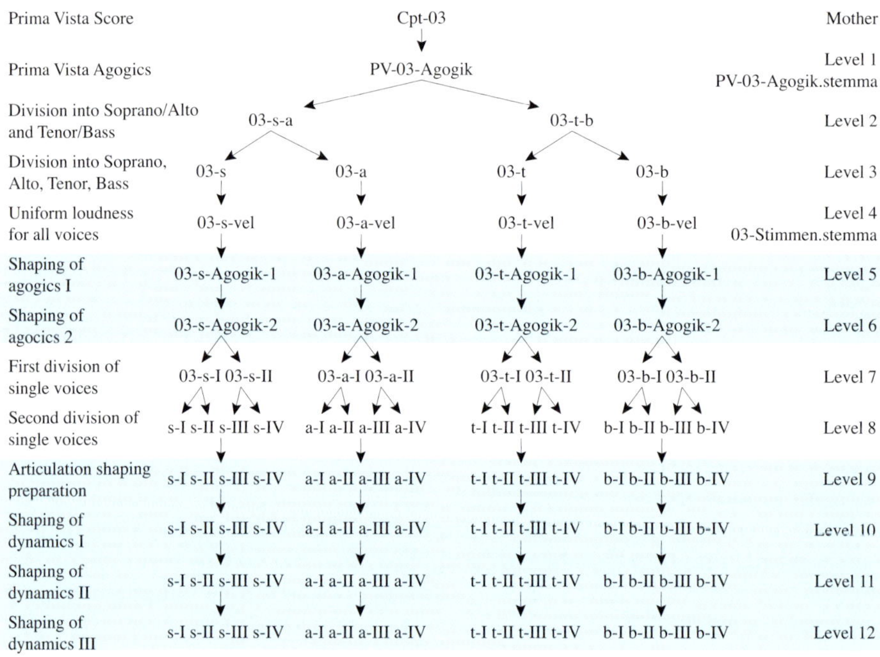

Fig. 22.10. The stemma of the third parcours.

- what is the sound of the analytical structure?
- can the sounding analytical structure yield a musically reasonable performance?

and two contrary performance strategies:

- the target-driven strategy,
- the experimental strategy.

In contrast to objective analytical approaches, when studying performance, subjective ingredients cannot be completely eliminated. They are present in their feedback with the performance result, while weights and intensity parameters in the WeightWatcher are determined, but they play a fairly reduced role.

From the first performance experiments, which have not been discussed in detail here, until the complete performance as described above, Stange-Elbe has known situations which demonstrated several problematic issues: It was not easy to eliminate the impression of an existing performance—in our case by Glenn Gould, say—and to stick strictly to what is written in the score; the performed version of the piece automatically resonates as a comparison while doing the performance work. This was the situation where Stange-Elbe started these experiments with the ambitious task of approaching an artistical and aesthetical performance as far as possible.

Therefore, the *target-driven* strategy was to a certain degree determined by the comparison with traditional human performances. Under these conditions, weights were applied and results were judged. This turned the tradition into an obstruction: It positioned the expected performance in the foreground and the shaping weight in the background.

Only the consequent questioning of the analytical structure and the systematic liberation from traditional performance expectations led to a performance strategy that positioned the analytical weights in the center of the investigation. This *experimental* strategy was coined by an as-unbiased-as-possible sounding realization of analytical structures, centered around the question of how a weight, when applied to a particular performance aspect, would sound. Within this procedure, it was possible to insert 'unheard' results, to admit purposed over-subscriptions in the sense of the 'still more clear,' whereas the question of whether an interpreter would play in this way turned out to be completely irrelevant.

From this point of departure, how a determined analytical structure would sound, the experimental approach to shaping a musically reasonable performance was sought. This qualitative determination of what is a "musically reasonable" performance is inevitably a subjective one which as such decides the subsequent steps toward the final performance. Much like the interpreter who puts up for discussion his provisionally final version while performing in concert—where in the last analysis it is more his personality than the musical performance which is judged—in computer-assisted performance, the subject who works with the performance workstation RUBATO® presents his results as a provisionally final contribution to the ongoing discussion.

When judging all these performances, one has to take into account that only metrical and motivic weights were applied and the effects of harmonic passages were not included in the shaping of performance (except of the motivations for the not-machine-made subdivisions from global to more local applications of weights in the third parcours). Furthermore, a certain economy in the choice of weights and their application was applied. In this sense, Stange-Elbe first had to check out which weights would involve what type of shaping consequences, and how the change of intensity parameters would influence the musical expressivity. It was only after this preliminary work that a systematic application of the weights and a partially purposed work with their intensity parameters became possible.

The portability of the presently described performance technique must be deduced from the compositional structure (a fugue in general and the thematic structure of the *Kunst der Fuge* in particular) as well as from the instrumental context. *In nuce* it can be said that such systematic statements are still premature. Many more analyses and performances would be necessary, but these can only be realized as soon as RUBATO® has become a common tool of musicology. Then the question can be asked whether general recipes can be stated that are valid beyond the limits of single compositions, or whether performance is rather bound to each individual composition.

The problem of historical context is undoubtedly a difficult one in view of systematic approaches. Should one use different analytical weights as a function of the historical situation (*Kunst der Fuge* requiring different weights/operators than *Träumerei*)? Or should one just use different operators with given weights? Could it be that at certain historical moments, the strong stress on weights' expressivity is more accepted than at other moments, where the interpretation is set more inside the listener's imagination?

Whatever is true for the transformation of the analytical structure in a scientific work targeting an artistically valid aesthetic performance, one should not forget about the elimination of (and nonetheless omnipresent) emotional and gestural aspects. The realization of a sonification of analytical structures during the interaction with the computer always bears a degree of emotionality, a phenomenon that should be taken into account as a kind of "uncertainty relation."

The judgment of the performance results took place in the same line as the judgment of a human performance, and the work with RUBATO® was also proposed as a provisionally final contribution to the work's discussion.

While describing the performance results, stress was put on a scientific analytic performance. The feedback to the analysis has a particular significance in that the conclusive character of a performance possibly could yield an analytical criterium. This implies an absolutely serious attitude toward analysis and no disclosure from emergent new aspects and innovative analytical ways of hearing.

Therefore, Stange-Elbe refrains from a discussion of subjects such as "prejudices against results which are produced by a machine", or "performance and the soul of music versus soulless performance machines". Instead, Stange-Elbe favors representations of procedures and performance strategies, the exemplary demonstration of connections between analyzed structures, performed results, and the attempt at a generalization of these insights in the form of a performance grammar in its dependency on the instrumental conditions.

23

Statistics

O sancta simplicitas!
Jan Hus (1370–1415)

This chapter deals with a special type of experiment in performance theory, since experiments are necessary to test the relevance of a theoretical approach to real performance. How can we know that such an approach is explaining what we are experiencing in performance? The question is quite hard, because it is difficult in music to distinguish the creative subjective aspect from the scientific objective one. There are essentially two types of experiments with an expressive performance theory:

- Construct synthetic performances and test their quality by psychometrical methods, as done by the KTH school, for example. This is the psychological approach. It is important, but it does not tell us how to construct performance tools except by trial and error. It just takes the subjectivity of the listeners as a variable and ponders it against the output of a performance machine.
- Take human performances and investigate their fitting quality with rationales of the theory, e.g. with analytical, gestural, or emotional rationales. This one also refers to the aesthetic human individual dimension, but it realizes it in the realm of performers—if possible even distinguished performers, such as Horowitz, Brendel, or Pollini. The comparison is not with synthetic performances but with rationales of performance. This is completely logical, since the performance's expressivity refers to those rationales. Therefore, these experiments should reveal correlations between performance and some rationale(s), and—in the limit—provide us with suggestions about the functional relation supporting such correlations.

In this chapter, we focus on the second method. This research was done in collaboration with statistician Jan Beran. The musical material we con-

G. Mazzola, *Musical Performance*, Computational Music Science,
DOI 10.1007/978-3-642-11838-8_23, © Springer-Verlag Berlin Heidelberg 2011

sidered was Schumann's *Träumerei* op.15/7, Webern's *Variationen für Klavier* op.27/II, the *canon cancricans* from Bach's *Musikalisches Opfer* BWV 1079, and Schumann's *Kuriose Geschichte* op.15/2. We have calculated metrical, motivic, and hormonic weights for all of these compositions.

The main task was then to transform this data into a format that was adequate for statistical processing. Since we were focusing on agogics, which had been measured by Bruno Repp for 28 famous performances, our analytical weights were all "boiled down" to functions of onset only. Therefore, we have taken the average values at a given onset for melodic and harmonic weights.

We should add here that Repp's measurements can not be done with much more precision and also regarding parameters other than time. The software Melodyne [105] editor (figure 23.1) is capable of transforming audio data to note data and, after an unavoidable amount of editing, into MIDI data. Therefore, the performance research is open to a huge repertory of historical recordings.

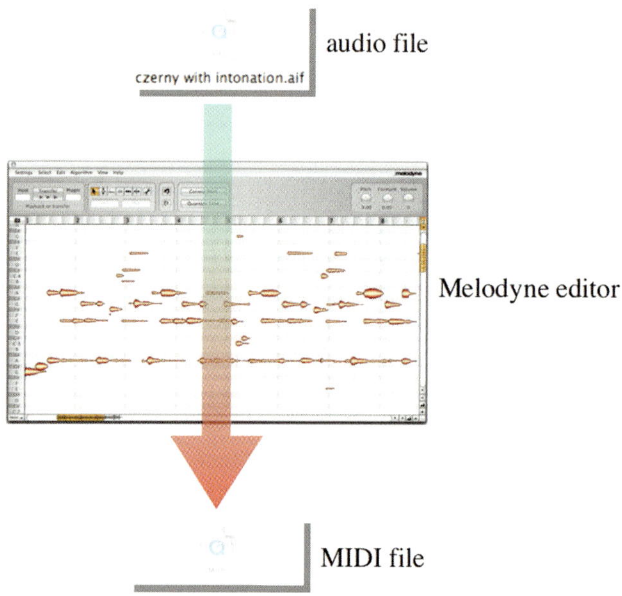

Fig. 23.1. The software Melodyne editor can transform polyphonic audio data into MIDI data and thereby opens research to performance analysis of historical recordings.

23.1 Hierarchical Decomposition of Weights

The statistically relevant decomposition of weights runs as follows: We start from a weight function $w(E)$ being defined for all onsets E, so it is a splined weight function, not just the discrete weight. The weight is not decomposed according to a Fourier procedure, because there is no reason to suppose that periodic weights should play any particular role in this context. Instead, we have chosen a hierarchical averaging procedure. Intuitively, this means that we start with a broad averaging of the weight, then deduct this from the weight and make a slightly less broad averaging, etc., thereby getting more and more local information represented on the finer averaging levels.

More precisely, we take a triangular support function

$$\begin{aligned} K(s) &= 1 - |s| \text{ for } s \in [-1, +1] \\ &= 0 \text{ else.} \end{aligned}$$

Given a sequence $(t_i)_{i=1,\ldots n}$ of times and a non-negative real number b, we next define the Naradaya-Watson kernel function by

$$K_b(t, t_i) = \frac{K(\frac{t-t_i}{b})}{\sum_{j=1}^{n} K(\frac{t-t_i}{b})}.$$

We then suppose given a time series of dimension k

$$(x_s(t_i))_{s=1,\ldots k}, i = 1, \ldots n.$$

The averaging formula then is this:

$$K_b x_s(t) = \sum_{i=1}^{n} K_b(t, t_i) x_s(t_i).$$

For $b = 0$, we have $K_0 x_s(t) = x_s(t)$.

The averaging process now works when we suppose that a decreasing sequence of bandwidths $b_1 > b_2 > \ldots b_m = 0$ is given. We first average according to b_1. This gives the new smoothed functions

$$x_{1,s} = K_{b_1} x_s.$$

We then proceed by induction. Suppose we have constructed smoothed functions $x_{1,s}, x_{2,s}, \ldots j-1$. Then we define the jth smoothed function by

$$x_{j,s} = K_{b_j}(x_s - \sum_{l=1}^{l=j-1} x_{l,s}).$$

In figure 23.2, we show the smoothing curves for a succession of bandwidths $8 > 4 > 2 > 1 > 0.5 > 0.1 > 0$ and *Träumerei*.

For our hierarchical smoothing process, we now start with the triangular support function

$$\begin{aligned} \hat{b}(s) &= 1 - |s|/b \text{ for } s \in [-b, +b] \\ &= 0 \text{ else.} \end{aligned}$$

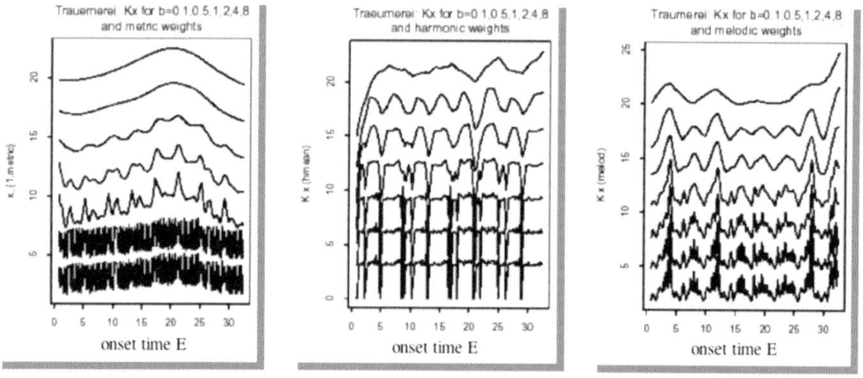

Fig. 23.2. Hierarchical smoothing curves for *Träumerei*—metric, harmonic, and melodic, from left to right—and bandwidths $8 > 4 > 2 > 1 > 0.5 > 0.1 > 0$.

and define the smoothed function for function f by

$$b \diamond f(E) = \int \hat{b}(t-E) \cdot f(t). \tag{23.1}$$

It averages f around E with weighted center E and *bandwidth* b. If this function is a weight, this means that the weight's analysis within the entire bandwidth neighborhood of a given onset is included instead of spiking the analysis to the singular onset. In the following process, this kernel smoothing process has been applied to a hierarchy of bandwidths, starting with $b = 4$ (= eight measures), then $b = 2$, then $b = 1$. The averaging process is taken to define successive remainder functions as follows:

$$f_1 = 4 \diamond f,\ f_2 = 2 \diamond (f - f_1),\ f_3 = 1 \diamond (f - f_1 - f_2),\ f_4 = f - f_1 - f_2 - f_3 \tag{23.2}$$

This means that the decomposition

$$x = x_1 + x_2 + x_3 + x_4 \tag{23.3}$$

for a smooth weight x defines a "spectrum" of that weight with respect to successively refined neighborhoods of its ambit.

Musically speaking, as already observed, this kernel smoothing process is completely natural. In fact, the kernel function alters the original time function $f(E)$ by a weighted integration of f-values in the kernel neighborhood of a given time E. This means that we now include the information about f from the neighboring times to make an analytical judgment. This latter is a well-known and common consideration in musical performance: The interpreter looks up a full neighborhood of a time point to derive what has to be played in that point. Moreover, the repeated application of the kernel smoothing process with increasingly narrowed neighborhoods is understood as a succession of a refinement in local analysis: First, the interpreter makes a coarse analysis over eight measures ($b = 4$), then he/she looks for the remainder $f - f_1$ and goes on with refined actions, if necessary.

This procedure is applied to the metric, melodic, and maximal and mean harmonic weights $x_{metric}, x_{melodic}, x_{hmax}, x_{hmean}$ and to their first and second derivatives $d_E x, d_E^2 x$. This gives the following list of a total of 48 *spectral analytical* functions:

$$\begin{array}{llll}
x_{metric,1} & x_{metric,2} & x_{metric,3} & x_{metric,4} \\
d_E x_{metric,1} & d_E x_{metric,2} & d_E x_{metric,3} & d_E x_{metric,4} \\
d_E^2 x_{metric,1} & d_E^2 x_{metric,2} & d_E^2 x_{metric,3} & d_E^2 x_{metric,4}
\end{array}$$

$$\begin{array}{llll}
x_{melodic,1} & x_{melodic,2} & x_{melodic,3} & x_{melodic,4} \\
d_E x_{melodic,1} & d_E x_{melodic,2} & d_E x_{melodic,3} & d_E x_{melodic,4} \\
d_E^2 x_{melodic,1} & d_E^2 x_{melodic,2} & d_E^2 x_{melodic,3} & d_E^2 x_{melodic,4}
\end{array}$$

$$\begin{array}{llll}
x_{hmax,1} & x_{hmax,2} & x_{hmax,3} & x_{hmax,4} \\
d_E x_{hmax,1} & d_E x_{hmax,2} & d_E x_{hmax,3} & d_E x_{hmax,4} \\
d_E^2 x_{hmax,1} & d_E^2 x_{hmax,2} & d_E^2 x_{hmax,3} & d_E^2 x_{hmax,4}
\end{array}$$

$$\begin{array}{llll}
x_{hmean,1} & x_{hmean,2} & x_{hmean,3} & x_{hmean,4} \\
d_E x_{hmean,1} & d_E x_{hmean,2} & d_E x_{hmean,3} & d_E x_{hmean,4} \\
d_E^2 x_{hmean,1} & d_E^2 x_{hmean,2} & d_E^2 x_{hmean,3} & d_E^2 x_{hmean,4}
\end{array}$$

For which *musical reasons* are these derivatives added to the analytical input data? The first derivatives measure the local change rate of analytical weights. Musically speaking, this is an expression of transitions from important to less important analytical weights (or vice versa), i.e., a transition from analytically meaningful points to less meaningful ones (or vice versa). This is crucial information to the interpreter: It means that he/she should change expressive shaping to communicate the ongoing structural drama. In the same vein, information about second derivatives is musically relevant because it lets the interpreter know that the ongoing structural drama is being inflected. Evidently, one could add higher derivatives, but we argue that an interpreter is already highly skilled if he/she can take care of all these functions and also observe different analytical aspects, from metrics to harmonics, simultaneously.

Besides these analytical input functions, we add three types of 'sight-reading' functions. They regard the following three primavista instances: ritardandi, suspensions, and fermatas. We omit these weights and refer to [84, chapter 44] for details. The entire spectral averaging procedure yields 58 functions of symbolic time E. Their vector, with all functions given a fixed order of coordinates, is denoted by $X(E) \in \mathbb{R}^{58}$.

Next, we look for a connection of this big analytical vector function to the tempo function found from Repp's analysis. We introduce this operator for $\omega \in \mathbb{R}^{58}$:

$$\Omega^X_\omega = (X, \omega),$$

the scalar product of ω and the analytical vector X. This means that for every onset E, we have $\Omega^X_\omega(E) = (X(E), \omega)$. Recapitulating the meaning of the analytical vector X, we are dealing with a second-order differential operator that we call a "Beran operator" since it was introduced by Jan Beran in [6].

On this basis, the central question of the following is whether tempo curves T of the *Träumerei* as they appear in the context measured by Repp in [111] may be approximated via Ω^X_ω by an appropriate choice of the shaping vector ω. The main result of this approach states that there is strong statistical evidence for the equation

$$\ln(T) = \Omega^X_\omega + C \tag{23.4}$$

for the given analytical vector X, a suitable shaping vector ω, and a constant C.

This means that the 58 coefficients of the shaping vector ω are random variables and that we prove a significant statistical correlation—in the mathematical form described by the Beran operator—between a certain subset of the analytical vector X and tempo as it is measured for the 28 performances by Repp.

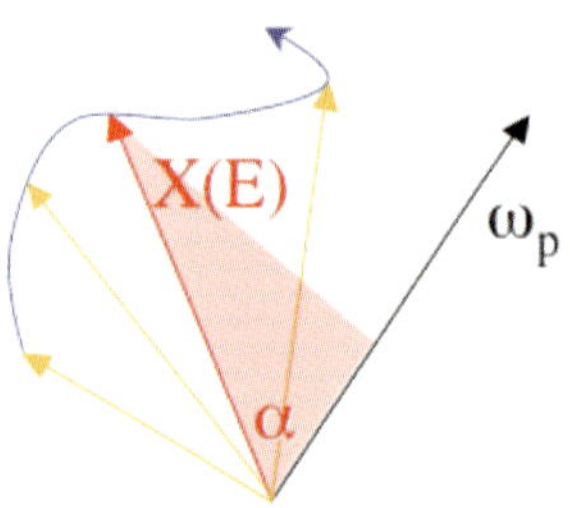

Fig. 23.3. The Beran operator uses the scalar product of the shaping vector ω and the analytical vector X at time E.

Observe that the formula in 23.4 uses the logarithm of tempo, a remarkable fact, which fits in the general fact that logarithms are important for cognitive processes, pitch and loudness being the classical cases. Moreover, taking the logarithm of tempo turns the set of all tempi into a real vector space: $\ln(T_1) + \ln(T_2) = \ln(T_1 T_2)$, and $\lambda \ln(T) = \ln(T^\lambda)$ are reasonable operations of tempo curves!

This being so, the hypothesis to be verified as a statistical statement is that for each p of those famous 28 artists playing *Träumerei*, the measurements of Repp enable a vector $\omega_p \in \mathbb{R}^{58}$ such that

$$\ln(T_p) = \Omega^X_{\omega_p} + C$$

is well approximated (figure 23.3).

This Beran operator formula is strongly supported by the present data set. Moreover, it can be shown that a small number of weights is already significant for the overall effect.

The main statistical conclusions from the analysis can be summarized as follows:

- There is a clear association between metric, melodic, and harmonic weights and the tempo.
- The exact relationship between the analytic weights and an individual tempo curve is very complex. However, a large part of the complexity can be covered by our model.
- Commonalities and diversities among tempo curves may be characterized by a relatively small number of curves. There is in principle no unique way of attributing features of the tempo to exactly one cause (harmonic, metric, or melodic analysis). Which curves need to be used depends partially on which of the three analyses (harmonic, metric, melodic) has 'priority.' However, there seems to be a small number of canonical curves that are essentially independent of the priorities and which determine a large part of the commonality and diversity among tempo curves. Natural clusters can be defined.
- There is a natural way of reducing an individual tempo curve to a series of simplified tempo curves containing an increasing number of features.

The results here are closely related to Repp's work [111]. Repp applied principal component analysis to the 28 tempo curves. One of his main results is that Cortot and Horowitz appear to represent two extreme types of performances. Thus, in a heuristic way, Repp suggested classifying the performances according to their factor loadings into a Cortot and a Horowitz cluster, respectively. Repp's Horowitz and Cortot clusters are confirmed.

Part V

Inverse Performance

24

What Is Inverse Performance Theory?

Never trust the artist. Trust the tale.
The proper function of a critic is to
save the tale from the artist who created it.
David Herbert Lawrence (1885–1930)

Inverse Performance Theory is the study of the reconstruction of the performance process from a given performance. This is a completely natural situation, since we usually perceive music as listeners and have to imagine what the performance could express, what it conveys in its complex signification process. Whereas performance theory mainly focuses on direct performance, i.e. the construction of performance map $\wp$ and the investigation of such models (such as the statistical investigations with Beran, see chapter 23), we now encounter the inverse problem which is one degree more difficult than the former, namely the theory of reconstruction of performance processes from a given performance. This deals with the variety of possibilities in the fiber $\wp^{-1}(P)$ of a given performance (figure 24.1). It is the famous fiber problem, and we call it *inverse performance theory*. It deals with the question:

Given a (collection of) performance(s) P, which types of and how many performance transformations $\wp$ and rationales can we have that generate that (collection of) performance(s)?

Beyond the technical challenge of this question (find performance fields, stemmata, operators, etc.), we are facing a very practical topic, namely the status of music critique. In fact, music critique is a special type of applied inverse performance theory insofar as the critic investigates (or should do so...) the rationales of a given performance in the light of the score, other performances, knowledge and prejudices about the admissible/admitted rationales of performance, and also a number of individual/ideosyncratic 'noise,' including ignorance and personal performance (on that night, the critic had a bad restaurant experience, the honorary for that critique was too low, the weather

G. Mazzola, *Musical Performance*, Computational Music Science,
DOI 10.1007/978-3-642-11838-8_24, © Springer-Verlag Berlin Heidelberg 2011

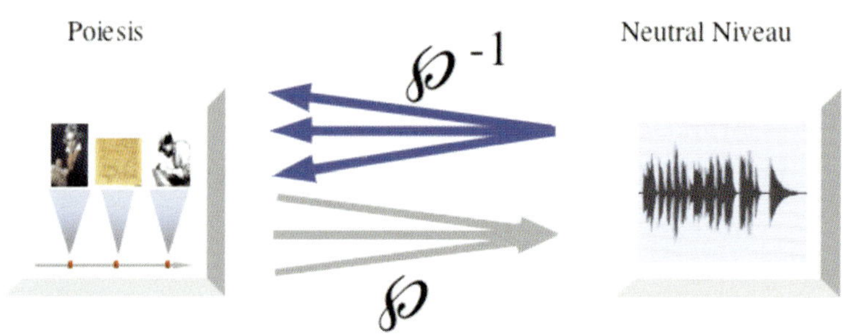

Fig. 24.1. Direct and inverse performance theory relate poiesis and neutral niveau in opposed directions.

was bad, a love affair had failed, and the neighbor in the concert was uninterruptedly coughing).

24.1 Technical Aspects

We first focus on the more technical side in order to get an insight about the complexity of the problem.

24.1.1 Reconstruction of the Performance Map from Recording Data

This is the so-called score-performance field matching problem. It has two parts:

24.1.1.1 Matching Algorithms

First, one needs so-called *matching algorithms*, which relate events of performed notes of a given score to their generating notes. This is difficult because

- simultaneous notes (chords) are often not played at the same time, not as an error, but as an effect of expressive performance.
- the score notes are not one-to-one represented in performance, either because the symbolism is ambiguous (e.g. for trills), or by errors of omission or note inaccuracies.

In the Espresso Rubette (see [103]), we have chosen a top-down strategy for this problem: Identify sets of notes corresponding to sets of events, and then, having enough such 'charts,' their intersection will be a single note/event pair, see also [104]. This algorithm works real-time. For an overview of matching algorithms, see [25].

24.1.1.2 Construction of Performance Fields

When the relationship between notes and events is created, one must generate performance fields according to the now existing map $\wp$. However, observe that this map is not defined in a frame, but only on a discrete set of points. Accordingly, the performance field $\mathbf{Ts}$ to be defined is also only defined on the notes of the given score, not between them, and we shall have to interpolate the discrete field.

The field construction needs three ingredients:

- The basis of the tangent space in every note. Since we do not have any information about the image of arbitrary neighboring points of a note X, we need to take a basis consisting of vectors $X \to X_i$ from X to neighboring notes X_i. To keep the information as local as possible, we have to choose the X_i as close to X as possible, but also as orthogonal as possible to each other (regarding the differences) in order to prevent the Jacobians from being singular.
- When these bases are gathered, the map applies and the image vectors are calculated, and then, the inverse Jacobians[1] are calculated and applied to the diagonal vector Δ.
- When this is all calculated, field interpolation is applied, and we get a real performance vector field.

The final problem of this reconstruction is visualization! We want to see the field, not only have its mathematical representation, which is too abstract. For one thing, it is n-dimensional, such as the piano space $\mathbb{R}^{EHLD}$ with four dimensions. To tackle this problem, we have to reduce the number of visible dimensions for a score representation to two, on a computer screen, say. So one can choose two dimensions of the score space, such as EH, and then two dimensions for the performance field, e.g. ED. The 2D visualization can also be done more intuitively by giving the vector direction a color from the color circle, and the length of the vector an intensity value. One example of such a color-based visualization was shown in figure 10.4 in section 10.3.

24.1.2 Reconstruction of the Stemma from the Performance Map

Once we have such a performance map/field, the question arises about the rationales and their mechanisms that may have led to such a performance. In this generality, the question has an infinity of answers, mostly intractable by lack of theory and mathematical complexity.

We have therefore investigated analytical rationales and relatively simple stemmata. We also have tried to take care of Beran's insight about the interaction of future, past, and present information in the shaping of performance at

[1] The Jacobian of a differentiable map f is the matrix $J(f) = (\partial_i f/\partial^j)$ of its partial derivatives.

a given moment of time. These were the semantics of his hierarchical smoothing method: to gather information about the past and future analytical values in order to define the averaged value at a given moment. Moreover, deeper mathematical investigations[2] about possible rationales have shown that if we are too generic about the possibilities, all performance rationales are in some sense generically equivalent, see [84, section 46.2].

Consequently, we have studied more concrete situations that are musically reasonable and then tried to understand what are the possible solutions in this situation.

Generally speaking, we are given a certain model M of expressive performance, including a bunch of analytical weights, a stemma, and corresponding operators (with their system parameters and weight inputs). We then suppose that the output performance $P = P_M(\text{parameters})$ (essentially the fields on the stemma's leaves) is given, and we ask for the fiber $P_M^{-1}(P)$, i.e., the set of all parameters yielding P with the given model P_M. This is the mathematical setup for inverse performance theory:

How many explanations for a given performance P are possible given a specific model P_M of understanding?

[2] Using modern Algebraic Geometry, this is work led by Roberto Ferretti.

25

The Technical Setup

To see a World in a Grain of Sand,
And a Heaven in a Wild Flower,
Hold Infinity in the palm of your hand,
And Eternity in an hour.
William Blake (1757–1827)

In this chapter, we want to make the idea explained in chapter 24 more precise in an explicit technical sense. Of course, this must rely on a concrete model of performance, one which is much more explicit than the generic setup for statistics, although statistics are a very strong argument for the construction of analytical rationales built upon weights and operators. We refer to figure 25.1 for the following discussion.

So this is the background structure leading to a given performance output: We identify the output by the set of performance fields, which are defined on the leaves of a stemma. They are the four framed bottom fields shown in our figure. This performance output stems from the input field of the primary mother shown on top of the stemma. This input field is usually just the diagonal default field.

The primary mother field is then altered by an operator shown by the blue circle arrow. This could be a primavista operator. Then, the resulting field is restricted to the three shown subframes. They might be defined by right hand, left hand, or some temporal delimitation.

After those restrictions, we get fields on daughters, as shown in our graphics. On each of these daughters, we have one operator of Lie type (this is a general assumption in view of the generic nature of Lie-type fields, as shown in section 17.1). The more important feature here is the interaction between different daughters of one and the same mother: sisters' family life. This interaction is precisely what Beran's averaging process was expressing! So we have to introduce operators that transfer information between sisters. These

G. Mazzola, *Musical Performance*, Computational Music Science,
DOI 10.1007/978-3-642-11838-8_25, © Springer-Verlag Berlin Heidelberg 2011

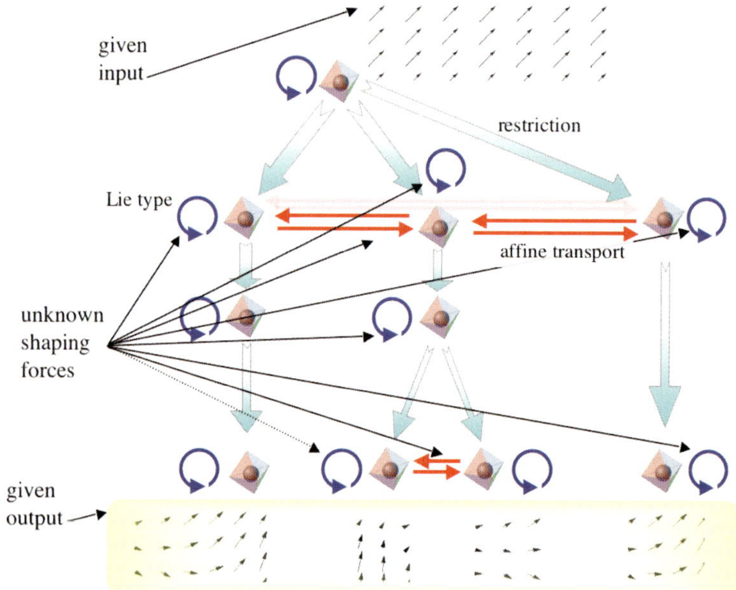

Fig. 25.1. The stemmatic display of operators and their family members under quasi-sexual propagation of LPSs.

operators take the field from one given LPS and transfer it to a sister frame by a unique affine morphism. All sister fields are then added when their affinely transported versions are available on a given frame. This process is visualized in figure 25.2. The summing of the sisters' transported fields is however also weighted according to a matrix $W^M = (W^M_{ij})$ of weights W^M_{ij} for sister i to sister j in the family of one mother M.

What is the meaning of such a weight? Suppose that the daughters of mother M are indexed according to increasing onset interval. So if $i < j$, then daughter i is a portion played earlier than daughter j, and the quantity W^M_{ij} measures the influence of daughter i on daughter's j expressive shaping. This is a causal influence, while the other case, $i > j$ is a final influence: The earlier daughter is influenced by the later daughter. The case $i = j$ means autocorrelation of daughter i.

This variety of values W^M for sister acts is very large, so we decided to restrict the shape of such a matrix to a variety defined by four real parameters: $causalStart$, $causalEnd$, $finalStart$, $finalEnd$. There is still a large variety of shapes that can be given to these matrixes; see figure 25.3 for nine examples of such matrixes that we represent as 'flying carpets.' The causal extremum is to the left, and the final extremum is to the right of each carpet surface.

With this setup of the stemma, we have a large number of system variables: four for the flying carpet of every mother, and then all the operators' parameters. Therefore, the inverse problem in this context is that we first

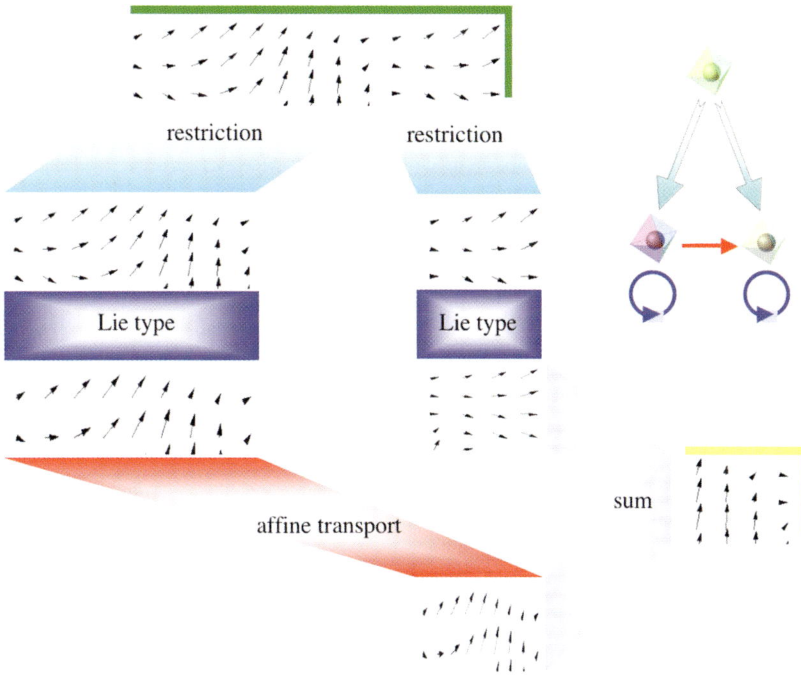

Fig. 25.2. The process of interaction between sister performance fields.

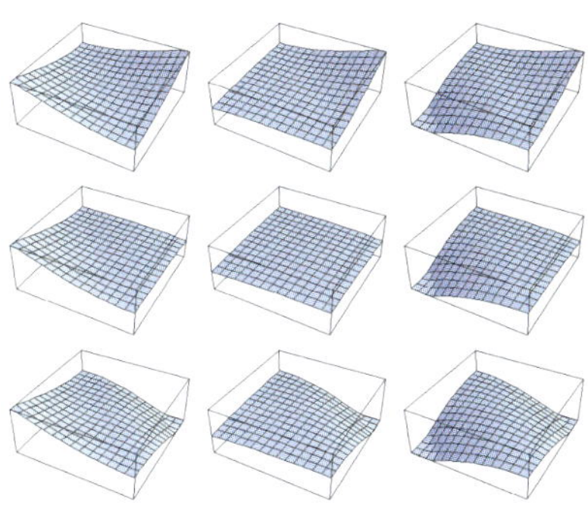

Fig. 25.3. Nine shapes of the interaction matrixes defined by four parameters, $causalStart, causalEnd, finalStart, finalEnd$.

have to choose one big stemmatic inheritance scheme with all its mothers and daughters as nodes. We then have to choose the Lie operators, together with their individual sets of weights in every node, and then we are ready to talk about numerical variables: the carpet parameters and the Lie operator parameters. Figure 25.4 shows the variety of interpretations generated by the present setup of stemmatic structures. For every output data, we have a fiber over

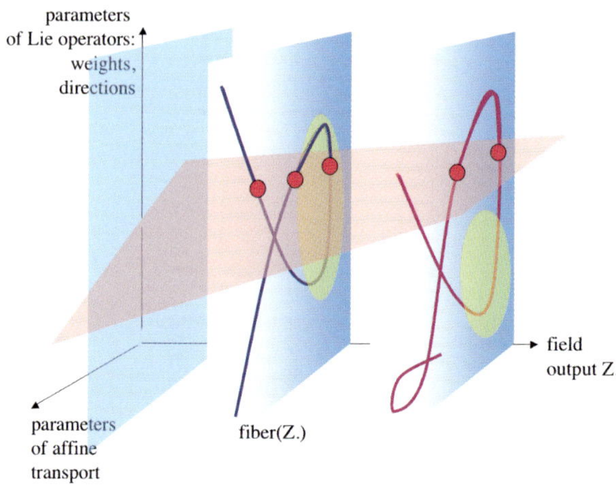

Fig. 25.4. The variety of parameters for given performances is shown in the space, which is spanned by the field output, the affine transport parameters, and the Lie operators' parameters with their weights. For every output data, we have a fiber over that data—this is drawn with a vertical blue plane, the parameters for that output being shown as a curve through the plane. The red surface represents the flying carpet-driven subvariety, with its relatively small number of points (in red) crossing the carpet-space and the fiber curve.

that data—this is drawn with a vertical blue plane, the parameters for that output being shown as a curve through the blue plane. The red surface represents the flying carpet-driven subvariety, with its relatively small number of points (in red) crossing the carpet-space and the fiber curve. Roberto Ferretti has shown [84, section 46.2] that generically these fibers are isomorphic, so it is necessary to restrict them to reasonably fine subvarieties in order to obtain characteristically different interpretative varieties.

26

Schumann's *Träumerei*: Argerich vs. Horowitz

Your mind must control, but you must have heart...
Give your feeling free.
Vladimir Horowitz

We want to apply the inverse theory described in chapter 25 to a concrete case: The agogical tempo fields as measured by Bruno Repp [111] from recordings by Marta Argerich (ARG in Repp's List, example ♫ 25) and Vladimir Horowitz (HO1 in Repp's List, example ♫ 26) of Schumann's *Träumerei*.

The stemmatic architecture is this: We start with the total piece on top, then split this total into four periods A, A', B, A''. Each period is finally split into its eight measures:

$$R_{A1}, R_{A2}, \dots R_{A8}, R_{A'1}, \dots R_{A'8}, R_{B1}, \dots R_{B8}, R_{A''1}, \dots R_{A''8}.$$

Each of the daughters inherits the restriction to the daughter's onset time frame of the global (boiled-down) motivic weight. The weight is then used as a tempo operator to stretch the local tempi. Moreover, the local tempi are altered by the scaled neighboring tempi (of the sisters), with a scaling by correlation coefficients from "flying carpets" controlling all sisters' linear correlations. There are five carpets: one from the primary mother to the period sisters, and one for each descent from a period to its eight measures. Each carpet has four parameters ($causalStart$, $causalEnd$, $finalStart$, $finalEnd$). Together with the original constant tempo dT, we have 21 parameters.

Local tempi are just restricted to the daughters and then processed via these operators. The 32 measure leaves have 32 constant tempi, which are the average tempi from the Repp measurements. A calculation of local minima (near solutions) of these equations yields this image (see [84, section 46.4.1] for details):

G. Mazzola, *Musical Performance*, Computational Music Science,
DOI 10.1007/978-3-642-11838-8_26, © Springer-Verlag Berlin Heidelberg 2011

Result for the period level:

- In the interperiod coherence, Argerich is more final than Horowitz, whereas the causal level is more pronounced by Horowitz.

Result for the bar level:

- Horowitz plays the first period with pronounced causal and final coherence, whereas the causal coherence decreases to a very low level toward the end of the piece.
- The repetition A' of the first period A shows a 'relaxation of coherence,' which may be justified by the repetitive situation.
- The development section B slightly increases the nal character.
- For Horowitz, the recapitulation seems to be quite 'tired': The causal character is very low, the nal character is decreased.
- For Argerich, the first period has a less coherent ambitus than with Horowitz.
- In contrast to Horowitz, the final coherence of Argerich increases as the piece goes on.
- The development and the recapitulation are pronouncedly final. The development and the recapitulation show a consciousness of the end of the piece that is absent with Horowitz.
- In other words, Argerich's recapitulation is 'prospective' and not 'retrospective.'

Fig. 26.1. Argerich's and Horowitz's performances of agogics are compared with respect to their background stemma for agogical shaping. Both artists define one point in the fiber. The parameters of such a point give us information about the coherence of time shaping. The gross result is that Argerich is more globally aware of the piece, whereas Horowitz is more focused on the local features.

So the difference between these two performance strategies means that Argerich is more coherent, integrating the entire piece's information when playing a specific spot, whereas Horowitz is more focused on that spot without caring too much about what has been played or what will be played. This does not mean that Argerich's strategy is musically better since the integration of all information for the performance of a specific spot means that every spot is played the same way! This is not generating a differentiated performance. We believe that a mixture of local and global information would yield the most thought-out result.

27

Rethinking Music Critique

At the age of 76, Florence Foster Jenkins finally yielded to public demand and performed at Carnegie Hall on October 25, 1944. So anticipated was the performance that tickets for the event sold out weeks in advance.
Jenkins died a month later, seemingly also because of the destroying critique.

27.1 Boiling Down Infinity

We have learned in previous chapters that performance conveys a doubly infinite message: the infinity of interpretative perspectives as they are realized in music analysis, gesturality, and emotive expressivity, and the variety of performative shaping expressed in the infinitesimal vocabulary of performance fields.

Critics are very probably not aware of such a variety of backgrounds that may produce concrete performances. In particular with respect to (analytical) interpretation, they preferredly stick to the traditional canon of how the structure of a composition should be viewed and interpreted or analyzed. Of course, it is not clear whether music critics should be cognizant of possibly new interpretations, but once they have gone into their business, a creative dealing with analytical problems should be mandatory.

One may understand that this is not necessary *ante rem*, but after the event, a re-reading of the text should be considered, be it only for comparative handling of the present performance: Could it be that the artist discovered a new interpretation of the given text? In practice, the selection of an analytical interpretation (in the best case, auto-incompetent critics excluded...) is just a matter of limitations of time, energy, and interest, besides ignorance of the infinite variety of interpretations.

G. Mazzola, *Musical Performance*, Computational Music Science,
DOI 10.1007/978-3-642-11838-8_27, © Springer-Verlag Berlin Heidelberg 2011

As to the infinity of performance nuances, this is beyond the vocabulary of music critics and it is also beyond the present measurement technology for such data: In a common concert, no performance field reconstruction is feasible. So critics are *nolens volens* limited to describing performance by use of common language expressions ("elegant diminuendo blended by a mysterious pedaling cloud..."), which beyond their imprecision cannot relate rhetorical expression to semiotic expression.

So is feuilletonism inevitable? Or rather: Is such a bad feuilletonism inevitable? Is it necessary to play the game of a unique "best" performance whose expression has to move along unreflected paths of prejudices? The alternative would be to embed one's judgment in the potential infinity of analytical interpretation and expressive performance. And to keep this embedding omnipresent in the critical discourse. We argue that the most precious role of a music critic would be that of putting the infinity of perspectives of a musical work into evidence in every concert or CD review. These would be the crucial points:

- Infinity of analytical interpretations,
- infinity of expressive performances,
- infinity of correlations between expressive rationales and expressive rhetorics (shaping of the performance field and other stemmatic details).

Our discourse is not about bad or good quality in these specifications, in essence, the only quality is to teach us something about the work in question and about the relativity of each perspective. Suggesting a boiled-down finitistic or even unicorned view of art is a destructive reduction and hinders every understanding or progress in the arts.

27.2 Glenn Gould's Politically Incorrect Performance

Glenn Gould's performance of classical works from Bach to Webern is a testbed for a valid music critique. His also technically spectacular performances have evoked strong reactions that unveil a number of deficiencies in common critique styles.

Whereas Gould's Bach performances may be non-conformist but still acceptable and adequate for Bach's compositions, his performances of Beethoven's sonatas are beyond the supportable deviation from common taste. The famous critic Joachim Kaiser has described in [58] the most famous "mis-performance" of a Beethoven sonata on the example of Gould's presentation of op.57 *Appassionata*, hear example ♫ 1:

> *Bei Goulds Wiedergabe des allegro assai dürfte es sich um die verrückteste, eigensinnigste Darstellung handeln, die jemals ein Pianist einem Bethoven-Satz hat angedeihen lassen; und das will etwas heißen. Gould hält es für richtig, demonstrativ langweilig und gelangweit den*

> *Kopfsatz so zu bieten, als ob ein Beethoven-Verächter seinen Plattenspieler nur mit halber Geschwindigkeit ablaufen ließe. Tranig langsam, langweilig und gelangweilt, die Triller während des pp im Schneckentempo, während der Fortissimo-Stellen etwas rascher, quält sich die Musik vorbei. Man meint, der Pianist imitiere ein Kind, das mit erfrorenen Fingern die Appassionata vom Blatt spiele. Nur selten vergißt er dabei, daß er ja vergessen machen wollte, der genialische Glenn Gould zu sein.*

This critique is strongly based upon the commonly accepted reading of the Beethoven text as a passionate message that calls for temperament and stormy dynamics in performance, and not for analytical cool vivisection of such a vital piece of literature. In Kaiser's characterization, Gould's production is like a "child with frozen fingers in a sight-reading performance." Here, the different and aberrant performance is incorrect, even forbidden. It is a norm that the politically incorrect Gould violated and thus made the sonata ridiculous; Kaiser even comments that the sonata "remains silent" when confronted with such a misreading.

The basic hypothesis behind such an outrageous indignation is that Kaiser knows what and when and how the sonata (which is personalized here) would have communicated, and that crazy Gould just destroyed that known and accepted message. Kaiser invokes an installed performance grammar, which requires a passionate forte seventh degree cascade towards the piano on the dominant in measures 14-16 of allegro assai. Instead, Gould descends like a noble, bored lady and snobbishly sits down on the boring dominant fermata. No passion whatsoever.

The same, even more ridiculous deformation can be observed in Gould's performance of sonata op.106, *Hammerklavier*, see example ♫ 27. This case is even worse because one just thinks that Gould did not understand a single word of the text, that he simply was too ignorant for the performance task.

What happened? And why is Gould's Bach so much more successful? Evidently, Gould's microscopic performance method works for Bach, but not for Beethoven. Why does this microscopic view fascinate and illuminate Bach's work whereas it virtually kills Beethoven's sonatas? The point is that in Beethoven's work, there is an inbuilt performance grammar that is not engraved in the score but stems from the performance tradition as such—an oral tradition so to speak, an element of rhetorical communication that transcends written code. Instead, Gould reads the same code from the Bach and from the Beethoven scores and effectively demonstrates that there is a huge defect in Beethoven's written code; it is quite trivial, at least locally, and the written script is simply boring.

Gould has effectively given a quasi-mathematical demonstration that the same performance strategy cannot be applied to Bach and to Beethoven. The same analytical insight and the same rhetorical shaping yield completely different results for these composers. To me, this is a sensational lesson to teach a characteristic difference between Bach and Beethoven.

This is very clear in the descending seventh passage on measure 14, which runs on semiquavers after a triggering triplet of quavers at the end of measure 13. Gould effectively takes the double temporal rate of the semiquavers with respect to the quaver triplet, without any tempo increase, without any dynamic profiling, just letting us see the anatomy of this triadic descent structure. The common reading [139] of this passage is that of an explosion:

> *Die Explosion (a tempo, Auftakt zu T.14) erfolgt im niedersausenden Dominant-Arpeggio und f-Sextakkord (T.15), wird aber sogleich abgedämpft durch einen C-Dur-Sextakkord, p, T.16.*

With Gould, there is no explosion, just the written text, cleanly played but antagonistic to any such musical drama of which an explosion would testify. The common reading classifies this sonata as a musical drama and asks interpreters to integrate this semantic into their performance. Gould plays the *Appassionata* minus the commonly implied drama. The question here is whether this dramatic character is implicit in the score structure or whether it is an external determinant that has been added by historical standards—which Gould filters away to lay bare what he believes is a poor structural essence [45].

Let us therefore analyze the specific performative shape of the passage in question. To begin, its agogics is profiled against the temporal neighborhood, i.e., not only is the indication "a tempo" valid from the last three quavers of measure 13, but in measure 14, the resumed tempo is again increased. The dramatic performance contains an increase of tempo, and within that level, also an increase of tempo toward the middle of the descent. Further, the dynamics is not only the *forte* at the end of measure 13, but the target tones of each descending intervallic movement of the descent are played louder, maybe to a *ff* or *sf*. As a whole, this descent (with its added ascending tail in measure 15) is not simply a musical structure, but more an explosive *gesture* whose very beginning goes to the top pitch, falls down, and bounces back to the dominant fermata. This is not a written rationale, but it is a semantic unit that can easily be deduced if gestural semantics is to be included in the performance shaping. It has become evident from recent research that gesture is crucial to the understanding of Beethoven's compositions, see [47] and [93].

So Gould's experiment would demonstrate that Beethoven requires gestural rationales beyond analytical ones. Meaning that Beethoven's compositions have a performative added value of gestural nature that is far from dominant in Bach's architectural music. Observe, however, that this gestural character is not on the level of the interpreter's gestures, but it is a rationale in the performance grammar, a semiotic layer that is added to the score system. Summarizing, Gould's politically incorrect performance withdraws from the common dramatizing approach and gives us a unique insight about Beethoven. This does not however imply that Bach has no gestural subtext, but if it had one, then it would be a completely different one, more like a puppet's gesture—

the puppet being Glenn Gould—when being manipulated by God, the supreme puppeteer, if I am allowed to draw a very theatrical image.

Part VI

Epilogue

28

Summary of Performance Theory

A hard beginning maketh a good ending.

Performance theory is a part of music theory in its larger understanding. It is not classical music theory, which unfortunately and implicitly focuses on harmony, a miniature part of theoretical topics in music (comprising theory of motives, of rhythms, of tunings, of physical modeling of sound, orchestration, composition, algorithmic composition, representation, etc.).

Present performance theory deals with the transformation of a symbolic score into a physical sounding event set, therefore it does not (yet) deal with musical performance that is not based upon a score. For example, it does not deal with the improvisatory creation of music without scores or with scores that need essential creative competence beyond the reading of fragmentary scores, such as lead sheets in jazz.

Performance theory has two main concerns: structure and expression.

Structure theory deals with the precise and complete description of the structure of performance transformations, score $\rightarrow$ embodied sound: *What* is performance? We are aware that the level of embodied sounds is a wide field, since sound embodiment can strongly focus on the body, the gestural utterance in music, and less on sound as acoustical patterns. However, most of the present theory focuses on sound. This is not to downsize the gestural embodiment; we simply do not know enough to date.

The description of structure of performance includes a *performance cell*, a minimal set of structural components that enable performance:

- the symbolic kernel (the notes),
- a region in the kernel's parameter space, called frame of the performance,
- the initial set, a collection of events in the frame, where performance is predefined; the latter is called the initial performance.

G. Mazzola, *Musical Performance*, Computational Music Science,
DOI 10.1007/978-3-642-11838-8_28, © Springer-Verlag Berlin Heidelberg 2011

- Finally, we need a vector field on the frame, the performance field, which defines the performance transformation with target space being the space of physical parameters defined for the notes in the symbolic kernel.

A realistic performance is defined by a system of performance cells, which are connected by projections of parameter spaces. Such performance hierarchies build the complete information needed to perform the given symbolic notes. Performance hierarchies are used as structural components in the construction of performances from expressive data.

Expressive theory deals with the content-based aspect of performance. There is a message that is transmitted to the audience, which answers the question of *why* performance is shaped. This relates to the semantic dimension of music, namely the fact that score-based music communicates meaning. This is not the most general case, because music might be a gestural utterance, which does not communicate given meaning but produces it in the making, if meaning is addressed at all.

Expressive theory relates (roughly speaking) to three specifications of contents. First, on the psychological reality: emotions. Second, on the physical reality: gestures. Third, on the symbolic reality: analysis. *The main problem of performance theory is the shaping of performance structure as a function of these contents.* This is about rhetorics, the shaping of expression to convey contents in the best possible way. So the general scheme is that we are given any such contents and then should know how to shape performance, i.e. a performance hierarchy, in order to communicate that content. The instances that shape performance by a given content are called operators. So the general formula is Performance = Function(Contents, Operators).

While it seems difficult to deal scientifically with the rhetorical shaping of emotional and gestural contents, the analytical rhetorics have reached a detailed level of theory. The theory works on the principle that analytical processes of rhythmical, motivic, or harmonic nature yield results that can be fed into operators, which in turn shape performance.

Performance theory had its first historical roots in the shift from music theory as a theory of abstract music to a theory of human production (as opposed to the divine perspective) in the sixteenth century. The experimental aspect of performance research goes back to the eighteenth century, when the first performance recording machines were built, essentially to document inprovisation. There have been two threads of performance theory: philosophical abstract theories about performance and empirical research dealing with recording and simulation of performance. Since the Swedish research at the Kungliga Tekniska Hgskolan (KTH) in the early 1980s, the philosophical and empirical threads have been united and can now be presented as those two standard parts, theory and experiment, of any exact science that deals with nature, be it the human or the material nature.

Consequently, performance theory has been implemented also in software such that its concerns can be tested on the empirical, experimental level. Besides the KTH software Director Musices, we have discussed the performance software RUBATO® developed at the Computer Science Department of the University of Zürich and at the Computer Science Department of the TU Berlin in the last decade of the twentieth century. RUBATO® is built on the modular principle that analysis is separated from performance operators. The price thereof is that analytical results must all be delivered by weight functions. The operators all use such weights. There are operators acting on the symbolic kernel, on the physical output, and on the performance fields.

We have discussed a number of case studies of such experimentally constructed performances. We have also discussed the statistical arguments for connecting analytical facts to the shaping of performance.

29

Future Developments

May the dreams of your past be the reality of your future.

The prospects of future developments in performance theory are many. They are mainly split into two main threads: theory and experiment. Perhaps one should rename performance theory and call it *Performance Science*, splitting into theory and experiments.

On the theoretical level, future developments should focus on the development of gesturally and emotionally driven operators. It is by no means clear here how to formally include these rationales. And when they are conceived, how would an operator act on them? Is it reasonable to ask that the gestural and emotional inputs to be casted to weights? This would enable us to apply gestures and emotions to given operators. But the feasibility of such a casting principle is not clear.

Shaping performance is the big problem of investigating possible operators. We have a class of Lie operators, but it is not clear whether this covers all interesting cases. All the more because the Lie type could lose its importance if non-weight-based input comes from gestural, emotional, or even symbolic rationales.

On the experimental side, we need strong devices for information gathering of performances. It seems that we still do not have very good "matcher" algorithms to connect scores to MIDI-recorded performances. We also need programs and hardware to represent those performance fields and to learn about their applicability in the education of performers.

The application of performance science seems to me to be a very important motivation for future developments. We have to learn to teach this science and to apply it to performers' education, and also to the art of listening to music.

It would also be a positive development to start a discourse about the philosophical aspect of performance, since the deep semiotical, rhetorical, and social impact of performance is an important issue in music science. This likely

G. Mazzola, *Musical Performance*, Computational Music Science,
DOI 10.1007/978-3-642-11838-8_29, © Springer-Verlag Berlin Heidelberg 2011

would have consequences for the culture of music critique, and also for the culture of hearing music, which is still a difficult issue.

Part VII

References, Music Examples, Index

References

1. Adorno Th W: Der getreue Korrepetitor (1963). Gesammelte Schriften, Bd. 15, Suhrkamp, Frankfurt am Main 1976
2. Adorno Th W: Zu einer Theorie der musikalischen Reproduktion. Suhrkamp, Frankfurt/M. 2001
3. Aebersold J: VOLUME 1—"HOW TO PLAY JAZZ & IMPROVISE". Jamey Aebersold Jazz 2000
4. Agawu V K: Playing with Signs. Princeton University Press, Princeton 1991
5. Bars I: Extra Dimensions in Space and Time (Multiversal Journeys). Springer, Heidelberg 2009
6. Beran J and Mazzola G: Timing Microstructure in Schumann's *Träumerei* as an Expression of Harmony, Rhythm, and Motivic Structure in Music Performance. Computers and Mathematics with Applications, Vol. 39, Issue 5/6, 99-130, 2000
7. Bowden M: THE SECRET HISTORY OF MILES DAVIS IN THE 1980s. `http://www.jazzitude.com/milesdavis_complete1.htm`
8. Bresin R and A Friberg: Synthesis and decoding of emotionally expressive music performance. IEEE SMC '99 Conference Proceedings, Vol. 4 pp. 317 - 322, 1999
9. Buteau Ch: Motivic Topologies and Their Signification in Musical Motivic Analysis. Masters Thesis, U Laval/Québec 1998
10. Buteau Ch: Motivic Spaces of Scores Through RUBATO's MeloTopRUBETTE. In: Perspectives in Mathematical and Computational Music Theory, Mazzola G, Th Noll, E Lluis-Puebla (Eds.): Verlag epOs-Music, Osnabrück, pp. 330-342, 2004
11. Biok H-R: Zur Intonationsbeurteilung kontextbezogener sukzessiver Intervalle. Bosse, Regensburg 1975
12. Camurri A et al.: EyesWeb: Toward Gesture and Affect Recognition in Interactive Dance and Music Systems. Computer Music Journal, 24:1, pp. 5769, 2000
13. Cavaillès J: Méthode axiomatique et formalisme, Hermann 1938
14. Châtelet G: Figuring Space. Kluwer 2000
15. Cherlin M: Schoenberg's Musical Imagination. Cambridge University Press 2007
16. Clynes M: Secrets of life in music. In: Analytica, Studies in the description and analysis of music in honour of Ingmar Bengtsson, Royal Swedish Academy of Music, No 47 pp. 3-15, 1985
17. Czerny C: Klavierschule. 1840

G. Mazzola, *Musical Performance*, Computational Music Science,
DOI 10.1007/978-3-642-11838-8, © Springer-Verlag Berlin Heidelberg 2011

18. Crelle A L: Einiges über musicalischen Audruck und Vortrag. 1823
19. Csikszentmihalyi M: The Psychology of Optimal Experience. Harper and Row, New York 1990
20. Dahlhaus C et al.: Neues Handbuch der Musikwissenschaft, Bd. 1-13: Athenaion and Laaber, Laaber 1980-1993
21. Danuser H et al. (eds.): Neues Handbuch der Musikwissenschaft, Bd. 11: Interpretation. Laaber, Laaber 1992
22. Miles D: The Complete Concert 1964: My Funny Valentine + Four and More (Live). Columbia 1964/Sony 1992
23. Deleuze G: Francis Bacon. La logique de la sensation. Éditions de la Différence, Paris 1981
24. Desain P and H Honing: The Quantization of Musical Time: A Connectionist Approach. Computer Music Journal Vol. 13 (3), pp. 56-66, 1989
25. Desain P: Make Me a Match: An Evaluation of Different Approaches to Score-Performance. CMJ 24:1 2000
26. Desain P and H Honing: Tempo Curves Considered Harmful. In: Music Mind and Machine: Studies in Artificial Intelligence, Thesis Publishers, Amsterdam 1992
27. Eimert H and H U Humpert: Das Lexikon der elektronischen Musik. Regensburg: Bosse Verlag 1973
28. Evans B: Portrait in Jazz—Bill Evans Trio. LP, Riverside 1960
29. Epstein D: Shaping Time: Music, the Brain, and Performance. Schirmer Books, New York 1995
30. Euler L: De harmoniae veris principiis per speculum musicum representatis (1773). In: Opera Omnia, Ser. III, Vol. 1 (Ed. Bernoulli, E et al.). Teubner, Stuttgart 1926
31. Fallows D: "Tempo and expression marks." In: Grove Music Online. Oxford Music Online, `http://www.oxfordmusiconline.com`
32. Finscher L: Studien zur Geschichte des Streichquartetts. Bärenreiter, Kassel 1974
33. Fleischer A: Eine Analyse theoretischer Konzepte der Harmonielehre mit Hilfe des Computers. Magisterarbeit, MWS, HU Berlin 1996
34. Frank H: RUBATO® Broadcast. ORF2: Modern Times, Jan. 10, 1997
35. Friberg A: Generative Rules for Music Performance: A Formal Description of a Rule System. Computer Music Journal, Vol. 15, No. 2, 1991
36. Friberg A: A fuzzy analyzer of emotional expression in music performance and body motion. In: J Sundberg and B Brunson (Eds.): Proceedings of Music and Music Science, Stockholm, October 28-30, 2004
37. Friberg A and R Bresin and J Sundberg: Overview of the KTH rule system for music performance. Advances in Experimental Psychology, special issue on Music Performance, 2006
38. Friberg A: `www.speech.kth.se/music/performance/download/dm-download.html`
39. Friberg A: `www.speech.kth.se/music/performance/performance_rules.html`
40. Friberg A: `www.speech.kth.se/music/performance/Texts/final_ritard.htm`
41. Gabrielsson A: Timing in Music Performance and Its Relation to Music Experience. In: Sloboda J (Ed.): Generative Processes in Music. Calderon Press, Oxford 1988
42. Gabrielsson A: Expressive Intention and Performance. In: Steinberg R (ed.): Music and the Mind Machine. Springer, Berlin et al. 1995

43. Goldin-Meadows S: Hearing Gesture: How Our Hands Help Us Think. Harvard University Press 2003
44. Gottschewski H: Tempohierarchien. Musiktheorie, Heft 2, 1993
45. Gould G: The Glenn Gould Reader. Alfred A. Knopf, New York 1984
46. Hardenberg F von (Novalis): Heinrich von Ofterdingen. 1802
47. Hatten R: Interpreting Musical Gestures, Topics, and Tropes. Indiana University Press 2004
48. Hawking S: A Brief History of Time: From the Big Bang to Black Holes. Bantam Books, New York 1988
49. Hecquet S and A Prokhoris: Fabriques de la danse. PUF, Paris 2007
50. Hegel G W F: Vorlesungen über die Aesthetik (1835). Werke in 20 Bänden mit Registerband: 13: Vorlesungen ber die Aesthetik I: BD 13, suhrkamp, Berlin 1986
51. Hentoff N: Liner notes to Coltrane's last album "Expression". Impulse AS-9120, New York 1967
52. Hjelmslev L: La Stratification du Langage. Minuit, Paris 1954
53. Jackendoff R and Lerdahl F: A Generative Theory of Tonal Music. MIT Press, Cambridge MA, 1983
54. Juslin P N: Cue utilization in communication of emotion in music performance: Relating performance to perception. Journal of Experimental Psychology: Human Perception and Performance, Vol. 26, pp. 1797-1813, 2000
55. Juslin P N, A Friberg, and R Bresin: Toward a computational model of expression in performance: The GERM model. Musicae Scientiae special issue 2001-2002, pp. 63-122, 2002
56. Juslin P N and J A Sloboda: Handbook of Music and Emotion (Affective Science). Oxford University Press, 2010
57. Kant I: Kritik der reinen Vernunft. Meiner, Hamburg 1956
58. Kaiser J: Beethovens 32 Klaviersonaten und ihre Interpreten. Fischer, Frankfurt/Main 1979
59. Kelley J L: General Topology. Van Nostrand, Princeton et al. 1955
60. Kendon A: Gesture: Visible Action as Utterance. Cambridge University Press 2004
61. Koelsch S et al.: Children Processing Music: Electric Brain Responses Reveal Musical Competence and Gender Differences. Journal of Cognitive Neuroscience Vol. 15(5), 2003
62. Kopiez R: Aspekte der Performanceforschung. In: de la Motte H: Handbuch der Musikpsychologie. Laaber, Laaber 1996
63. Kopiez R and Langner J: Entwurf einer neuen Methode der Performanceanalyse auf Grundlage einer Theorie oszillierender Systeme. In: Behne K-E and de la Motte H (eds.): J.buch der D. Ges. für Musikpsychologie. **12** (1995), Wilhelmshaven 1996
64. Kronman U and Sundberg J: Is the Musical Ritard an Allusion to Physical Motion? In: Gabrielsson A (ed.): Action and Perception in Rhythm and Meter. Bubl. of the Royal Swedish Acad. of Sci. 55, Stockholm 1987
65. Langer S: Feeling and Form, Routledge and Kegan Paul, London 1953
66. Langner J, Kopiez R, Feiten B: Perception and Representation of Multiple Tempo Hierarchies in Musical Performance and Composition: Perspectives from a New Theoretical Approach. In: Kopiez R and Auhagen W (eds.): Controlling Creative Processes in Music. Peter Lang, Frankfurt am Main et al. 1998

67. Lazarus, R S: Emotion and Adaptation. Oxford University Press, New York 1991
Proceedings of the Int. Conf. on Music and Dance, Roehampton University 2005
68. Lewin D: Generalized Musical Intervals and Transformations (1987). Cambridge University Press 2007
69. Lewis C I: Mind and the World Order. Dover 1991
70. Lipscomb S and G Mazzola: Appreciation of complex musical sound: The influence of a visual performance component. Proceedings of the SMPC 2007
71. Listenius N: Musica... 1537
72. Liszt F: De la situation des artistes (1835). Liszt's Gesammelte Schriften, II. Band
73. Livingstone S R et al.: Changing Musical Emotion: A Computational Rule System for Modifying Score and Performance. Computer Music Journal, 34:1, pp. 4164, 2010
74. Mazzola G et al.: A Symmetry-Oriented Mathematical Model of Classical Counterpoint and Related Neurophysiological Investigations by Depth-EEG. In: Hargittai I (ed.): Symmetry II, CAMWA, Pergamon, New York 1989
75. Mazzola G: Geometrie der Töne. Birkhäuser, Basel 1990
76. Mazzola G and E Hunziker: Ansichten eines Hirns. Birkhäuser, Basel 1990
77. Mazzola G: *presto* Software Manual. SToA music, Zürich 1989-1994
78. Mazzola G: *Synthesis*. SToA 1001.90, Zürich 1990
79. Mazzola G and Muzzulini D: Tempo- und Stimmungsfelder: Perspektiven künftiger Musikcomputer. In: Hesse H P (ed.): Mikrotöne III. Edition Helbling, Innsbruck 1990
80. Mazzola G and Zahorka O: The RUBATO Performance Workstation on NeXTSTEP. In: ICMA (ed.): Proceedings of the ICMC 94, San Francisco 1994
81. Mazzola G: Inverse Performance Theory. In: ICMA (ed.): Proceedings of the ICMC 95, San Francisco 1995
82. Mazzola G et al.: Analysis and Performance of a Dream. In: Sundberg J (Ed.): Proceedings of the 1995 Symposium on Musical Performance. KTH, Stockholm 1995
83. Mazzola G: Semiotic Aspects of Musicology: Semiotics of Music. In: Posner R et al. (Eds.): A Handbook on the Sign-Theoretic Foundations of Nature and Culture. Walter de Gruyter, Berlin and New York 1998
84. Mazzola G: The Topos of Music. Birkhäuser 2002
85. Mazzola G and S Müller: Constraint-based Shaping of Gestural Performance. Proceedings of the ICMC 03, ICMA 2003
86. Mazzola G and S Mller: The Extraction of Expressive Shaping in Performance. Computer Music Journal, Vol. 27(1), MIT Press, 2003
87. Mazzola G: Musical Interpretation via Vector Fields: A New Human Workstation on the NEXTSTEP Environment. In: Friberg A et al. (Eds.): SMAC 93 Proceedings of the Stockholm Music and Acoustics Conference 1993, KTH, Stockholm 1994
88. Mazzola G: La vérité du beau dans la musique. Delatour 2007
89. Mazzola G and M Andreatta: From a Categorical Point of View: K-nets as Limit Denotators. Perspectives of New Music, Vol. 44, No. 2, 2006
90. Mazzola G and Andreatta M: Diagrams, Gestures, and Formulas in Music. Journal of Mathematics and Music 2007, Vol. 1, no. 1, 2007.

91. Mazzola G and P B Cherlin: Flow, Gesture, and Spaces in Free Jazz—Towards a Theory of Collaboration. Springer Series Computational Music Science, Heidelberg et al. 2009
92. Mazzola G et al.: Topos Theory for a Creative Analysis of Boulez's Structures. In: Naimpally S, G Di Maio (Eds.): Quaderni di Matematica, Theory And Applications of Proximity, Nearness and Uniformity, Vol. 23, 2009
93. Mazzola G: Categorical Gestures, the Diamond Conjecture, Lewin's Question, and the Hammerklavier Sonata. Journal of Mathematics and Music Vol. 3, no. 1, 2009
94. `wwww.rubato.org`
95. McNeill D: Gesture and Thought. University of Chicago Press 2005
96. Merleau-Ponty M: Phénoménologie de la perception. Gallimard 1945
97. Milmeister G: The Rubato Composer Music Software: Component-Based Implementation of a Functorial Concept Architecture. Springer Series Computational Music Science, Heidelberg 2009
98. `http://ns1.foxtail.com/Tech/mohrReader.html`
99. Montagu J: "metronome." In: The Oxford Companion to Music, Alison Latham A (ed.). Oxford Music Online, `http:// www.oxfordmusiconline.com`
100. Moog R: MIDI, Musical Instrument Digital Interface. Audio Eng. Soc. 34, no. 5, 1986
101. `http://www.midi.org`
102. Müller S: Pianist's Hands: Synthesis of Musical Gestures. PhD Thesis, University of Zurich, Zurich 2004
103. Müller S and G Mazzola: Constraint-based Shaping of Gestural Performance. Proceedings of the ICMC 03, ICMA, Ann Arbor 2003
104. Müller S and G Mazzola: The Extraction of Expressive Shaping in Performance. Computer Music Journal, Vol. 27(1), MIT Press, 2003
105. `http://en.wikipedia.org/wiki/Celemony#Melodyne`
106. Thomas Noll Th et al.: Computer-Aided Representation, Analysis and Conveyance of Musical and Music-Theoretical Structures, KIT-Report 149, Final Report of the Interdisciplinary Research Group KIT-MaMuTh for Mathematical Music Theory at the Technical University Berlin (1998-2003) 2004
107. Petsche H et al.: EEG and Thinking: Power and Coherence Analysis of Cognitive Processes. Austrian Academy of Sciences, 1998
108. Pfeifer R and Ch Scheier. Understanding Intelligence. MIT Press 1999
109. Pfeifer R and J Bongard: How the Body Shapes the Way We Think: A New View of Intelligence. MIT Press 2006
110. Puttka M: Learning to Dance Means Learning to Think! In: Bläsing B, Puttke M, Schack Th. (Eds.): The Neurocognition of Dance. Psychology Press, Hove and New York 2010
111. Repp B: Diversity and Commonality in Music Performance: An Analysis of Timing Microstructure in Schumann's *Träumerei*. J. Acoustic Soc. Am. Vol. **92**, 1992
112. Repp B: The Sound of Two Hands Clapping: An Exploratory Study. J,. Acoust. Soc. Am. 81 (4), April 1987
113. Reti R: The Thematic Process in Music (1951). Greenwood Press, Westport 2nd ed. 1978
114. Riemann H: Der Ausdruck in der Musik. Leipzig, 1883
115. Riemann H: Musikalische Dynamik und Agogik. Leipzig 1884

116. Ries F: Biographische Notizen über L. van Beethoven (1838). New edition by Kalischer A Ch, 1906
117. Russell J A: A Circumplex Model of Affect. J. of Personality and Social Psychology, Vol. 39, pp. 1161-1178, 1980
118. www.rubato.org
119. Sachs C: Rhythm and tempo; a study in music history. Norton, New York 1953
120. Sachs K-J: Der Contrapunctus im 14. und 15. Jahrhundert. AMW, Franz Steiner, Wiesbaden 1974
121. Schilling G: Musikalische Dynamik oder die Lehre vom Vortrag in der Musik. 1843
122. Scholes P et al.: "tempo." In: The Oxford Companion to Music, Alison L (Ed.). Oxford Music Online, http://www.oxfordmusiconline.com.
123. Schönberg A: Harmonielehre (1911). Universal Edition, Wien 1966
124. Seipp R: Analyse von Interpretationen mit dem Welte-System: Aufgezeigt an der Agogik der Polonaise A-Dur von Chopin. Das Mechanische Musikinstrument vol. 12; issue 44, 1988
125. Sessions R: Questions About Music. W. W. Norton and Company, 1971
126. Slater E and A Meyer: A contribution to a pathography of the musicians: 1. Robert Schumann. Confinia Psychiastricia, 2: 65-94, 1959
127. Shepard R N and L A Cooper: Mental Images and Their Transformations. MIT Press 1982
128. Siron J: La partition intérieure. Outre mesure, Paris 1992
129. Sluming V et al.: Broca's Area Supports Enhanced Visuospatial Cognition in Orchestral Musicians. The Journal of Neuroscience, Vol. 27(14), pp. 3799-3806, April 4, 2007
130. Stange-Elbe J: Analyse- und Interpretationsperspektiven zu J.S. Bachs *Kunst der Fuge* mit Werkzeugen der objektorientierten Informationstechnologie. Habilitationsschrift (manuscript, avaliable in several German libraries), Osnabrück 2000
131. Sundberg J, A Friberg, L Frydén: Rules for automated performance of ensemble music. Contemporary Music Review, Vol. 3, Issue 1, pp. 89-109, 1989
132. Sundberg J: Music Performance Reseach. An Overview. In : Sundberg J, Nord L, Carlson R (eds.): Music Language, Speech and Brain. London 1991
133. Sylvester D: Interviews with Francis Bacon. Pantheon Books, New York 1975
134. Tanaka A, Tokui N, and Momeni A: Proceedings of ACM Multimedia, Facilitating Collective Musical Creativity, 2005
135. Thalmann F: Musical Composition with Grid Diagrams of Transformations. Master's Thesis, University of Bern, Bern 2007
136. Todd N P M: Towards a Cognitive Theory of Expression: The Performance and Perception of Rubato. Contemporary Music Review **4**, 1989
137. Todd N P M: The Dynamics of Dynamics: A Model of Musical Expression. J. Acoustic Soc. Am. Vol. **91**, pp. 3540-3550, 1992
138. Türk D G: Klavierschule. 1789
139. Uhde J: Beethovens Klaviermusik III. Reclam, Stuttgart 1974
140. Uhde J and R Wieland: Forschendes Üben. Bärenreiter 2002
141. Ungvary T and R Vertegaal: The Sentograph: Input Devices and the Communication of Bodily Expression. ICMA 1995
142. Valéry P: Cahiers I-IV (1894-1914). Celeyrette-Pietri N and Robinson-Valéry J (eds.), Gallimard, Paris 1987

143. Valéry P: Philosophie de la danse. Oeuvres, tome I, Vairété, "Théorie poétique et esthétique", pp. 1390-1403, Gallimard, 1956
144. Wagner R: Über das Dirigieren (1869) (Translation: On Conducting. Dover, 1989)
145. Wagner Ch and H von Karajan: Experimentelle Untersuchungen ber das Tempo. Oesterreichische Musikzeitschrift 29, 1974, S. 589-604
146. Wanderley M and M Battier (Eds.): Trends in Gestural Control of Music. IRCAM - Centre Georges Pompidou, Paris 2000
147. Wang C Ch: Effects of Some Aspects of Rhythm on Tempo Perception. Journal of Research in Music Education, Vol. 32, No. 3 pp. 169-176, 1984
148. Widmer G: Modeling the Rational Basis of Musical Expression. Computer Music Journal Vol. 18, 1994
149. Wieser H-G and Mazzola G: Musical consonances and dissonances: Are they distinguished independently by the right and left hippocampi? Neuropsychologia Vol. 24 (6), pp. 805-812, 1986
150. Wieser H-G and Mazzola G: EEG responses to music in limbic and auditory cortices. In: Engel J Jr, Ojemann G A, Lüders H O, Williamson P D (eds.): Fundamental mechanisms of human function. Raven, New York 1987
151. Wittgenstein L: Tractatus Logico Philosophicus = Logisch-Philosophische Abhandlung, Wilhelm Ostwald (ed.), Annalen der Naturphilosophie, 14, 1921

Music Examples

A good example is the best sermon

In this book, a lot of examples may be invoked to illustrate the theoretical and practical discourses. We have however only included those examples in the following list that we consider being not only of pedagogical value, but also crucial for understanding our thoughts. The list uses two kinds of references: 1) references to Internet documents, 2) references to original documents of published music, where the first kind is not available for technical or legal reasons. The second kind is marked by an asterisk (*). The examples are listed in the order they first appear in the book.

♫ 10, page 66
www.encyclospace.org/special/webern_equal.m4a
♫ 11, page 66
www.encyclospace.org/special/MaqamBayyatiRacy-Shaheen.mp3
♫ 12, page 66*
Alois Hába: Streichquartette Gesamtaufnahme. Bayer, 2007
♫ 13, page 195
www.encyclospace.org/special/czernyalone.aif
♫ 14, page 137*
Anton Webern: Complete String Trios & Quartets (Arditti quartet). Disques Montaigne, 2000
♫ 15, page 144
www.speech.kth.se/music/performance/Texts/final_ritard.htm
♫ 16, page 153
www.encyclospace.org/special/schumannLHtakefive.aiff
♫ 17, page 166*
Miles Davis: The Complete Concert 1964: My Funny Valentine + Four and More (Live). Columbia 1964/Sony 1992
♫ 18, page 195
www.encyclospace.org/special/chopin.aif
♫ 19, page 198
www.encyclospace.org/special/kufbeispiel.aif
♫ 20, page 198
www.encyclospace.org/special/Kuriose_G_Deadpan.aiff
♫ 21, page 198
www.encyclospace.org/special/Kuriose_G_Final.aiff
♫ 22, page 208
www.encyclospace.org/special/kufcp3_deadpan.aiff
♫ 23, page 208
www.encyclospace.org/special/kufcp3_finish.aiff
♫ 24, page 235*
Argerich, Marta: Kinderszenen, Kreisleriana. Deutsche Grammophon 1984
♫ 25, page 235*
Horowitz, Vladimir: Kinderszenen, Op. 15 Scenes from Childhood. Original recording 1974, Sony Classical/Legacy 2003
♫ 26, page 240*
Gould, Glenn: Ludwig van Beethoven, op. 57. Originally released 1970, Sony Classic, 2008
♫ 27, page 241*
Gould, Glenn: Piano Sonata No. 29 in B-flat Major, Op. 106 "Hammerklavier". Original recording 1970, Sony Classical 1993

Index

D

E

G

H

I